Table of Contents

Design and Organize Manufactured Products for Optimum Operational Availability   *Page 2*

Preface

Availability is an important metric used to assess the performance of repairable systems, accounting for both the Reliability (R) and Maintainability (M) properties of a component or system. There are different classifications of availability and different ways to calculate it.

The classification of Availability is somewhat flexible and is largely based on the types of downtimes used in the computation and on the relationship with time (i.e., the span of time to which the availability refers). As a result, there are a number of different classifications of availability, including:

> Instantaneous (or Point) Availability
> Average Uptime Availability (or Mean Availability)
> Steady State Availability
> Inherent Availability
> Achieved Availability
> Operational Availability

The design parameters generally include reliability, maintainability of the system. Reliability will be the result of manufacturing of the system where as maintainability is coming from operational, maintenance logistics, inventory management, Prognostic Health Management (PHM) and supply chain design of the services that system require. The general usage of the PHM philosophy of system design is geared toward outcome-based contracts. The prognostics approach is a more effective way to maintain desired operational availability. It allows reduced administrative and logistic delay times by anticipating upcoming failures and preparing the necessary parts in advance. In addition, the prognostics capability allows for intelligent maintenance replacing only those parts whose remaining lifetime has reached a critical value. Availability is the major factor of operational effectiveness along with performance of the system. Availability based contracts are not as complicated as performance based contract because the discussion over metrics and requirement is less obscure to define for customer and designer. Minimum required availability of complex system is a key factor of many distributed and repairable systems.

Operational Availability (Ao) is a measure of the "real" average availability over a period of time and includes all experienced sources of downtime, such as administrative downtime, logistic downtime, etc. The operational availability is the availability that the customer actually experiences.

Ao is essentially the posteriori availability based on actual events that happened to the system. The previously indicated availability classifications are a priori estimates based on models of the system failure and downtime distributions. In many cases, operational availability cannot be controlled by the manufacturer due to variation in location, resources and other factors that are the sole province of the end user of the product.

Optimization is an important tool in making decisions and in analyzing physical systems. In mathematical terms, an optimization is the task of finding the best solution from among the set of all feasible solutions.. The Operational Availability optimization requires minimizing the sum of Preventive and Corrective Maintenance costs. and includes the cost of lost system operation; those quantities costs are generally known as economic analyses

1.0 INTRODUCTION

1.1 General

This write-up addresses availability elaborating the concept of operational availability and its impact on system design, operational supportability and life cycle cost.. Operational Availability is a calculation of various supportability functions at the systems level. The desired result of performing these calculations, coincident with system design, is to provide fielded systems with greater capability for the Subject Matter Expert and enhanced support at the best possible value. Operational Availability (Ao) provides a method of predicting and assessing system performance and readiness during the acquisition process and then becomes the performance benchmark during Initial Operational Capability (IOC), deployment and operations/maintenance cycles. This writeup is a practical guide, providing several useful equations and checklists to assist a Program Manager to understand and use Ao is a useful metric in the design and support of a weapon system.

1.2 Availability A(t)

λ = Failure Rate (Failures/Hr) $\lambda = 1$ / MTBF

μ = Repair Rate (Repairs/Hr) $\mu = 1$ / MTTR

A Transition Matrix is created for Availability and Laplace transfom used to derive Availability Equations

---

Case 1: Availability of Single Unit with Repair in Use Time "t"

Instantaneous Availability is

$$A(t) = \frac{\mu}{\lambda+\mu} + \frac{\lambda}{\lambda+\mu} e^{-(\lambda+\mu)t}$$

State 1: The Unit is operating and thus available for use
State 0: The Unit has failed and is undergoing repair

Steady State Availability is $A_{SS} = A(\infty) = \dfrac{\mu}{\lambda+\mu}$

Case 2: Availability of Two Equal Units in Parallel with Multiple Repairs in Use Time "t"

Let s1 = - 2 ($\lambda + \mu$) and s2 = - ($\lambda + \mu$)

Instantaneous Availability is

$$A(t) = \frac{\mu^2 + 2\lambda\mu}{\lambda^2 + 2\lambda\mu + \mu^2} + \frac{2\lambda^2}{s1s2(s1-s2)}(s1e^{s2t} - s2e^{s1t})$$

State 2: Both Units are operable and operating
State 1: One Unit is operable and operating, the other unit has failed and being repared
State 0: Both Units have failed, system has failed, and repairs are being made on both units

Steady State Availability is $A_{SS} = A(\infty) = \dfrac{\mu^2 + 2\lambda\mu}{\lambda^2 + 2\lambda\mu + \mu^2}$

1.3 Mission Availability

Mission Availability is the expected availability for a given mission period. Mathematically this is expressed as $A_m (t_2 - t_1) = \frac{1}{(t2 - t1)} \int_{t1}^{t2} A(t)\, dt$ Usually t_1 is considered as zero

$$Am (t2 - t1) = \frac{\mu}{(\lambda + \mu)} + \frac{\lambda}{(\lambda + \mu)^2} (e^{-(\lambda+\mu)t1} - e^{-(\lambda+\mu)t2})$$

Steady State Availability is the portion of Up Time expected for continous operation. Mathematically this is expressed as

$$A_{SS} = \lim_{t \to \infty} A(t)$$

1.4 Understanding Operational Availaility A_o

A_O is a probability function of reliability, maintainability and supportability components. Very simply, this equation is:

$$A_o = \frac{System\ Up\ Time}{System\ Up\ Time + System\ Down\ Time}$$

Total Time has two sub-factors, UP Time and DOWN Time. UP Time is the time a system is operational between failures. DOWN Time is the time the system is not operational

Now, what does this statement mean? First, Operational Availability is a supportability goal; the satisfaction of this goal will be determined during the system's design/test, and then the goal becomes a metric for evaluating operational performance through-out the system life cycle. Operational Availability is the supportability calculation of the equipment/system (hardware & software) in terms of predicted Reliability (R) called Mean Time Between Failure (MTBF) and predicted Maintainability (M) in terms of Mean Time To Repair (MTTR) and designed supportability, called Mean Logistics Delay Time (MLDT). As the hardware and software are designed (or selected in the case of Commercial Off The Shelf COTS), the logistics support system must also be designed (selected concurrently to meet program requirements. Figure 1.1 displays the interaction of the measures calculated to determine A_O.

The first calculations performed to generate an A_O are to determine inherent system reliability excluding consideration of support functions (e.g., re-supply, transportation, and repair); this metric is called Inherent Availability (A_i). The predicted R&M values are used in a basic A_i equation as shown in Figure 1-1.

After A_i has been determined, we now are ready to add the supportability calculation to consider logistics support system impacts on system performance. This is described as 'Mean Logistics Delay Time' (MLDT). Figure 1-1 helps to better understand the difference between A_i and A_O

The benefits of calculating Ai principally apply to the design and support of electronic systems. Ai calculations for mechanical and electrical systems are based on very predictable wear and tear experience. For example, a bearing will have a historical wear-out rate based on values for temperature, pressure and operating time. Using these type variables allows development of preventive maintenance schedules based on predictable failures. With this information, the supply chain can anticipate demand and procure and position spares and repair parts in anticipation of the wear-out. On the other hand, electronics components have random failure rates and failures can only be described as the probability of failure over a period of time.

Now let's look at where we obtain the information required to determine the variables shown in Figure 1-1. Reliability (R) is a probability function based on the actual physical components in the design and how often they randomly fail during a fixed time period. The equation is:

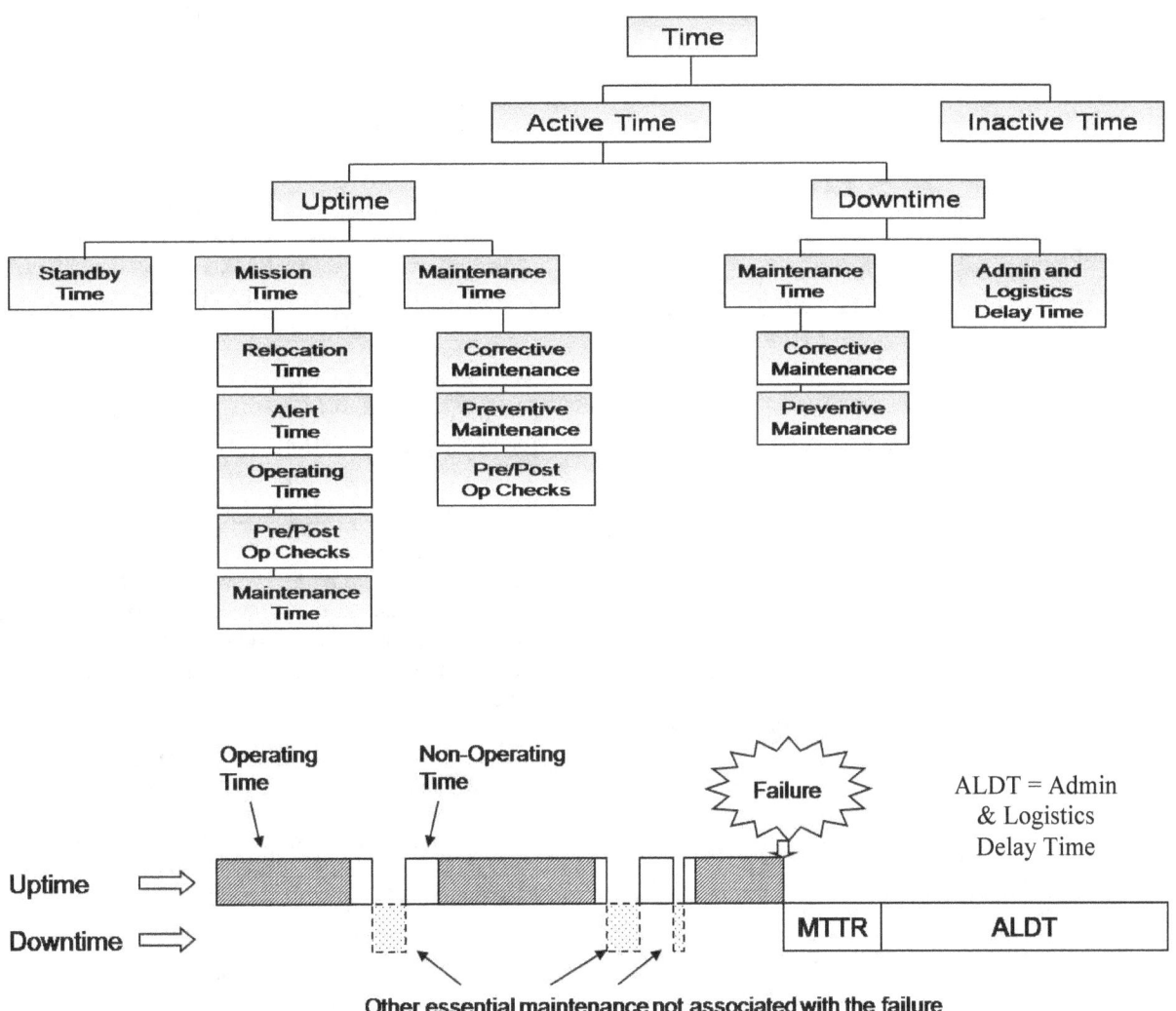

Categories of Time for Use in Defining A_O

Availability Types	Availability Equation	Availability Described

Inherent Availability A_i	$$A_i = \frac{MTBF}{MTBF + MTTR}$$	• Assures operation under stated conditions in an ideal customer service environment (no delays experienced while maintenance is being performed). It excludes:
		  ○ Preventive or scheduled maintenance (i.e., battery replacement, oil change, etc.
		  ○ Logistics delay times (i.e., filling out paperwork).
		
		• A_i is usually not specified as a field-measured requirement, since the customer service environment is rarely due to:
		  ○ Insufficient number of spare parts
		  ○ Long delays to obtain repair parts
		  ○ Inadequate training of repair personnel
		  ○ Excessive administrative requirements

Achieved Availability A_a	$$A_a = \frac{MTBM}{MTBM + MTTRactive}$$	• Similar to A_i, except that preventive and scheduled maintenance actions are factored into the uptime variable (MTBM). The corresponding preventive and scheduled maintenance times are included in the MTTRactive parameter. A_a is usually not specified as a field-measured requirement, since the downtime factor does not consider the routine logistics and administrative delays that occur during normal field conditions
Operational Availability A_o	$$A_O = \frac{MTBM}{MTBM + MDT}$$	• Extends the definition of A_i to include delays due to waiting for parts or processing paperwork in the mean downtime parameter (MDT). • A_O reflects the real-world operating environment, thereby making it the preferred and most readily available metric for assessing quantitative performance. • A_O is usually not specified as a manufacturer-controllable requirement without being accompanied by estimates of the logistics resources and administrative delays, induced failures, etc. which are government driven and beyond the manufacturers control.

MTBM = Mean Time Between Maintenance (A measure of reliability taking into account the maintenance policy), which is the total number of life units expended by a given time for, maintenance events (scheduled & unscheduled) performed on that item

MTBF = Mean Time Between Failure

$MTTR_{active}$ = Mean Time To Repair for corrective and preventive maintenance

MDT = Mean Down Time includes Mean Logistics Delay Time (MLDT), other delays and MTTR

Figure 1-1 : Variation of Availability Equation

1.5 Variables in Availability

$$R(t) = e^{-\lambda t}$$
$$R(t) = e^{-(\frac{t}{\eta})^{\beta}}$$

$$MTBF = \int_0^{\infty} R(t)dt$$

R being a decimal of less than one, e is the natural logarithm, λ (lambda) is the component failure rate, η is the Weibull Scale Parameter, β is the Weibull Shape Parameter and t is the time period over which the failures are tracked. Often Reliability is defined as MTBF. This is a simple concept based on the component failure rate λ over some time period. For example, if a component failure rate is 500 failures per million hours it follows that the reliability (MTBF) is equal to a million hours (test time) divided by 500 failures, which generates an MTBF of 2000 hours. Therefore, MTBF is the reciprocal of the failure rate. (not always)

Now that we have the reliability based on MTBF the next item to calculate is the MTTR. (Details in MIL-HDBK-470A Appendix D)

$$MTTR = \frac{\sum_{n=1}^{N} \lambda n \, Tn}{\sum_{n=1}^{N} \lambda n}$$

N = Number of replaceable items (RI)
λ_n = Failure Rate of the nth RI
T_n = Mean Repair time of the nth RI

This is the time it takes to remove interference, remove, replace and test the failed component, return the equipment to its original condition, and replace and retest any system/interference removed to get to the failed equipment. Next is MLDT, the cumulative time required by all logistics processes to support the requisite repair. MLDT may be a difficult factor to quantify because it includes parameters such as depot repair Turn Around Time (TAT), administrative delay time, supply response time and other factors that impact the maintenance/repair effort.

MLDT factors generally are combined measures that include Customer Wait Time (CWT) that is made up of three possible measures - Mean Supply Response Time (MSRT), Mean Outside Assistance Delay Time (MOADT) and Mean Administrative Delay Time (M$_{adm}$DT). Thus, the Ao equation can be restated as:

$$MLDT = MSRT + MOADT + M_{adm}DT$$

$$A_o = \frac{MTBM}{MTBM + MTTR + MSRT + MOADT + M_{adm}DT}$$

It is easy to see that because MTBM is both above and below the equation line, changes in its value have relatively limited impact on the Ao. MTTR, which is usually a small number (for electronic systems), also has minimal impact on the overall Ao value. The main driver of Ao is MLDT, which is often called the support system effectiveness measure. Changes in MLDT such as transportation times and depot TATs, typically have large values usually measured in days, weeks or months and, thus, have a major impact on the denominator for the calculation.

1.6 Availability (A_S) of a Series and Parallel System

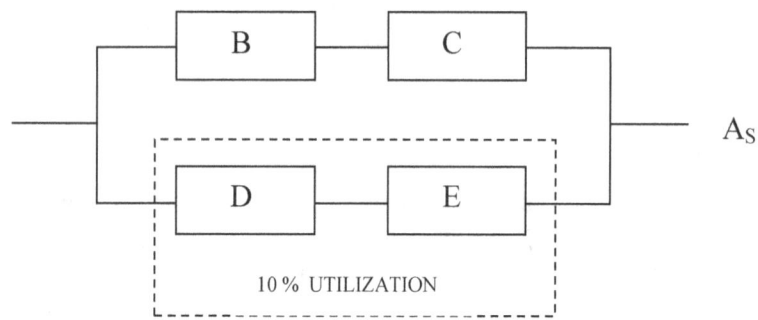

Say MTBF of each Sub-System is 1500 Hours
Say MTTR of each Sub-System is 30 Minutes
Say Logistics Delays are none

$A_B = A_C = 1500 / (1500 + 0.5) = 0.999667$
$A_D = A_E = 1500/0.1 / ((1500/0.1) + 0.5) = 0.999967$
$A_{BC} = (0.999667)^2 = 0.999334$
$A_{DE} = (0.999967)^2 = 0.999934$

$A_S = 1 - (1 - 0.999334)(1 - 0.999934) = 0.99999996$

1.6.1 Simplified Example Inserting Values for the Availability Variables:

MTBF of 1000 hours
MTTR of 3 hours MLDT of 3000 hours
$A_o = 1000 / (1000 + 3 + 3000) = 0.25$

To a fleet user, this A_O would be viewed as exceptionally poor and in all likelihood would not meet the needs of the mission

So where would we apply focus to improve the Ao value? Using this same example, let's look at a major weapon system capability improvement program that has a significant impact on the Fleet in terms of time to install and costs millions of dollars to implement. Let's say for this example that the improvement in Reliability (MTBF) increases by 30% (a very significant improvement in reliability) but does not improve the supportability factors of the system. Using this new MTBF figure (1300 hours) to recalculate Ao:

$A_o = 1300/ (1300 + 3 + 3000) = 0.30$

Fleet users will not notice this .05 increase in the A_O value even though the reliability has been improved by 30%. Therefore, a large increase in reliability has little measurable impact on A_O.

Now, if an investment is made in improving the supply chain (an action having minimum direct impact on the Fleet) and MLDT is reduced from 3000 to 1200 hours, the impact on A_O is more significant and results in an A_O value of 0.45:

$$A_o = 1000 / (1000 + 3 + 1200) = 0.45$$

This reduction in MLDT results in an 80% improvement in the Ao from 0.25 to 0.45. From this example, we clearly see that once a system is fielded, increasing the effectiveness of the logistics support pipeline is more effective than enhancing the system reliability. When MTTR is a large number, maintenance improvements/time reductions can also have a significant impact.

1.7 Use of Availability as a Conceptual Specification

For more than twentyfive years, Federal Aviation Administration (FAA) specifications have focused primarily on availability requirements in place of the more traditional reliability and maintainability requirements that preceded them. Availability requirements are useful at the highest levels of management. They provide a quantitative and consistent way of summarizing the need for continuity of Customer Services. They can facilitate the comparison and assessment of architectural alternatives by FAA headquarters system engineering personnel. They also bring a useful performance metric to analyses of operationally deployed systems and Life Cycle Cost tradeoffs. And, because availability includes all sources of downtime and reflects the perspective of system users, it is a good overall operational performance measure of the performance of fielded systems.

There are, however, significant problems with employing availability as a primary RMA requirement in contractual specifications. This operational performance measure combines equipment reliability and maintainability characteristics with operation and maintenance factors that are beyond the control of the contractor as well as outside of the temporal scope of the contract

The fundamental concept of availability implies that reliability and maintainability can be traded off. In other words, a one-hour interruption of a critical service that occurs annually is seen as equivalent to a 15-second interruption of the same service that occurs every couple of days — both scenarios provide approximately the same availability. It should be obvious that interruptions lasting a few seconds are unlikely to have a major impact on Air Traffic Control operations, while interruptions lasting an hour or more have the potential to significantly impact traffic flow and safety of operations. Contractors should not be permitted, however, to trade off reliability and maintainability arbitrarily to achieve a specific availability goal. Such tradeoffs have the potential to adversely impact Customer operations. They also allow a readily measured parameter such as recovery time to be traded off against an unrealistic and immeasurable reliability requirement following a logic such as: It may take two hours to recover from a failure, but it will be say 20000 hours between failures, so the availability is still acceptable.

As pointed out above, during system development, availability can only be predicted using highly artificial models. Following development, system availability is not easily measured during testing at the developer site. The fundamental statistical sampling limitations associated with high levels of reliability and availability are not the only problem. Availability cannot be measured directly; it can only be calculated from measurements of system downtime and the total operating time.

Deciding how much of the downtime associated with a failure should be included or excluded from calculations of availability is difficult and contentious. In an operational environment, matters are clear cut: simply dividing the time that a system is operational by the total calendar time yields the availability. In a test environment, however, adjustments are required for downtimes caused by things like administrative delays, lack of spares, and the like – factors that the contractor cannot control. Failure review boards are faced with the highly subjective process of deciding which failures are relevant and how much of the associated downtime to include.

1.8 Availability Improvement Considerations

Availability is the combined measure of Reliability and Maintainability. It should be maximized in the most cost effective manner. To achieve this the following requirements should me met

1. The designed-in failure rate should be minimized, and the MTBF or MTTF should be maximized

2. The designed-in repair rate should be maximized, and the MTTR should be minimized

3. As many maintenance actions as possible should be carried out while the equipment is running normally thus minimizing equipment downtime

4. If certain functions must be shutdown for maintenance, the time required for shutting down the equipment should be minimized

5. Should certain components require shutdowns for maintenance actions, these maintenance actions should be required as rarely as possible

6. Should certain maintenance actions require shutdown, the time needed for these actions should be minimized

7. If certain components or subsystems require shutdowns for maintenance actions, as few components as possible should be shut down

8. The time required for logistics should be minimized

9. The time required for administrative actions should be minimized

10. Very well written and explicitly illustrated startup and operating manuals should be prepared and made available to the users of the equipment and to the maintenance personnel.

11. Frequent and time consuming prescribed preventive maintenance actions should be minimized

12. Special effort must be expended to use qualified and well trained maintenance personnel. Their training should be updated as required as design changes and more modern equipment are introduced.

13. whenever the application of certain media, like lubricants, impregnants, etc., post a risk of damage to some components, these components should be protected and the risks involved in the application of these media should be clearly explained by the supplier of these media.

14. Maintenance actions which require the dismantling, moving and assembling of heavy components and equipment should be facilitated by the provisioning of special lift-off lugs and accessories

15. Frequently inspected, serviced, maintained and/or replaced components should be so located in the equipment that they are more accessible and easily visible

16. Servicing media like lubricants, impregnants, detergents, fuels, and other consumables should preferably be supplied automatically, and waste media should be removed automatically

17. Whenever possible automatic diagnostics for fault identification should be provided via failure indicating hardware and/or special minicomputers with the associated software

18. Specially designed and built-in automatic test and checkout equipment should be provided whenever feasible

19. The distributions of all equipment downtime categories should be determined and studied and those maintenance actions which contribute excessively to the overall equipment downtime should be singled out and their downtimes should be minimized

20. The distributions of the equipment downtimes resulting from the failure of key components should be studied. Those components contributing significantly to the overall equipment downtime should be singled out, and these components should be redesigned with lower failure rates and higher repair rates

1.8.1 Availability Workbench

Availability Work Bench (AWB) is the latest in a line of Isograph products *[https://www.isograph.com/about-us/]* to serve the Reliability and Maintenance community.. AWB integrates updated versions of the AvSim+ (Availability Simulation Software) and RCMCost (Reliability Centered Maintenance) products. These products have been used in industry since 1988. It also includes a brand-new life cycle cost analysis module. Availability Workbench provides a fully integrated environment for:

- Reliability Centered Maintenance. Developing and maintaining a Reliability Centered Maintenance (RCM) program to optimize your reliability and maintenance strategy.

- Availability Simulation. Performing full system availability predictions taking into account complex dependencies on spares and other resources

- Life Cycle Cost Analysis. Performing a Life Cycle Cost Analysis to calculate the expected costs of your system during its lifetime.

1.9 Improving Availability through Reliability-Centered Maintenance (RCM)

All facilities that are currently in operation require maintenance to continue to properly perform their functions and support their assigned missions. An effective and efficient maintenance program saves resources and maximizes availability. Reliability-Centered Maintenance (RCM) is an approach for developing an effective and efficient maintenance program based on the reliability characteristics of the constituent parts and subsystems, economics, and safety.

1.9.1 RCM Introduction.

Prior to the development of the RCM methodology, it was widely believed that everything had a "right" time for some form of preventive maintenance (PM), usually replacement or overhaul. Despite this commonly accepted view, the results indicated that in far too many instances, PM seemed to have no beneficial effects, and, in many cases, actually made things worse by providing more opportunity for maintenance-induced failures.

1.9.2 RCM Definitions.

The following definitions are commonly used in connection with RCM.

- RCM is a logical, structured framework for determining the optimum mix of applicable and effective maintenance activities needed to sustain the operational reliability of systems and equipment while ensuring their safe and economical operation and support.

- Maintenance is defined as those activities and actions that directly retain the proper operation of an item or restore that operation when it is interrupted by failure or some other anomaly. (Within the context of RCM, proper operation of an item means that the item can perform its intended function).

- Corrective maintenance (CM) is maintenance required to restore a failed item to proper operation. Restoration is accomplished by removing the failed item and replacing it with a new item, or by fixing the item by removing and replacing internal components or by some other repair action.

- Scheduled and condition-based preventive maintenance conducted to ensure safety, reduce the likelihood of operational failures, and obtain as much useful life as possible from an item.

- Preventative maintenance is the act of doing maintenance in anticipation of some failure. Recent data collection efforts have indicated that PM programs are ineffective on a number of typical system components. An effective PM program will not only identify components that have predictable failures, but identify those that are random to avoid waste of maintenance resources.

1.9.3 RCM Overview.

The RCM approach provides a logical way of determining if PM makes sense for a given item and, if so, selecting the appropriate type of PM. The approach is based on a number of factors.

- RCM seeks to preserve system or equipment function.

- RCM is more concerned on maintaining end system function than individual component function.

- Use reliability as the basis for decisions. The failure characteristics of the item in question must be understood to determine the efficacy of preventive maintenance.

- Consider safety first and then economics. Safety must always be preserved. When safety is not an issue, preventive maintenance must be justified on economic grounds.

- Acknowledge design limitations. Maintenance cannot improve the inherent reliability; it is dictated by design

- Treat RCM as a continuing process. The difference between the perceived and actual design life and failure characteristics is addressed through age (or life) exploration.

1.9.4 Preventive Maintenance.

RCM has changed the approach to preventive maintenance. The RCM concept has completely changed the way in which PM is viewed. It is now widely accepted that not all items benefit from PM, and it is often less expensive to allow an item to run to failure rather than to do PM.

1.9.5 Condition Monitoring and Analysis.

Some impending failures can be detected using some form of condition monitoring and analysis, a type of preventive maintenance. Condition monitoring is defined as periodically or continuously checking physical characteristics or operating parameters of an item. Based on analyzing the results of condition monitoring, a decision is made to either take no action or to replace or repair the item. Condition monitoring can be performed through inspection, or by monitoring performance or other parameters. *[http://irantpm.ir/wp-content/uploads/2012/02/17359.pdf]*

1.9.6 The RCM Objective

RCM has two primary objectives: to ensure safety through preventive maintenance actions, and, when safety is not a concern, preserve functionality in the most economical manner. Preventive maintenance is applicable only if it is both effective and economically viable. When safety is not a consideration and PM is either not effective or less economical than running to failure, only CM is required.

- PM can be effective only when there is a quantitative indication of an impending functional failure or indication of a hidden failure. That is, if reduced resistance to failure can be detected (potential failure) and there is a consistent or predictable interval between potential failure and functional failure, then PM is applicable.

- The costs incurred with any PM being considered for an item must be less than for running the item to failure (economic viability). The failure may have operational or non-operational consequences. The two categories of cost included in such a comparison for these two failure consequences are operational - the indirect economic loss as a result of failure and the direct cost of repair, and non-operational - the direct cost of repair.

1.9.7 A Product Can Fail in Two Basic Ways

First, it can fail to perform one or more of the functions for which it was designed. Such a failure is called a functional failure. Second, a product can fail in such a way that no function is impaired. The failure could be something as simple as a scratch or other damage of the finish of the product. Or it could be that one of two redundant items, only one of which is required for a given function, has failed.

1.9.8 The Three Categories of Failure Consequences

 Generally categories used in RCM analysis are Safety, Operational, and Economic. If a functional failure directly has an adverse affect on operating safety, the failure effect is categorized as Safety. When the failure does not adversely affect safety but prevents the end system from completing a mission, the failure is categorized as an Operational failure. When a functional failure does not adversely affect safety and does not adversely affect operational requirements, then the failure is said to have an Economic effect. The only penalty of such a failure is the cost to repair the failure.

1.10   Metrics

1.10.1   Metric Attributes

A metric differs from a measurement in that a metric is a composite of meaningful, quantifiable product or process attributes taken over time that communicate important information about quality, processes, technology, products, and/or resources. Measurements are simply the raw data from which metrics are calculated. A good metric is capable of reliably measuring a specific process repeatedly over time. The purpose of a metric is to measure change, regardless of whether that change is positive or negative. For example, when measuring the technical performance of a test article, the goal is usually to increase the performance of the article. If the present test of the article reflects a decrease in performance from the previous test, the data, process, and any changes made must be examined to determine the cause. Metrics help define problems by fostering process understanding and indicating when corrective action is required. The goal of metrics is to show a trend that results in action to improve the process.

For a metric to be meaningful, it must represent one or more cause-and-effect relationships that control the process being measured. Sometimes data may be difficult to measure or collect, but it is very valuable. Other times, data that is easily collected is meaningless. Ensuring that data value is worth the collection effort is essential to a good metric. Metrics should be reconsidered if the data do not represent cause-and-effect relationships, do not show a trend, or are not timely. In addition, output metrics are preferred over input metrics.

Many metrics, such as those relating to cost, schedule, and performance, can be used throughout the program's life cycle, while others may be tied to only one portion of the program. Choosing quality over quantity of metrics is a continuing challenge. To assist teams in assessing metrics, the following is a list of attributes generally associated with a good metric:

- Has value to the team members or is an attribute essential to customer satisfaction with the product

- Tells how well organizational goals and objectives are being met through processes and tasks

- Is simple, understandable, logical and can be used repeatedly

- Shows a trend

- Is unambiguously defined

- Uses data that is cost-effective to collect

- Allows for timely collection, analysis, and reporting of information

- Provides insight that drives appropriate action

1.10.2 Types of Metrics

Metrics are used at all levels of a program's structure to represent key measures mainly in the areas of cost, schedule, and performance. Metrics monitored by program Integrated Ptoduct Teams (IPTs) should be supported by other metrics monitored by sub-tier teams. Sub-tier teams should ensure that their metrics are aligned with the Program IPT's metrics in intent, language, and format. Sub-tier team metrics, however, should not be limited to those handed down from the program team. At the sub-tier level, additional metrics are often needed to accurately monitor the performance of the sub-tier team's product.

Many programs have a software component. Efforts in measurement for software development include Practical Software Measurement (PSM) by the Joint Logistics Commanders Joint Group on Systems Engineering. Their website has links to other software measurement sites. *The PSM website is at* http://www.psmsc.com

Three major categories of metrics are progress, product, and process. These categories are intended to assist teams in identifying the types of metrics they should be using.

1.10.3 Progress Metrics

Progress metrics are used to monitor the health and status of the program. They serve as alarms for adverse trends. These metrics must allow for the detection of adverse trends in sufficient time to permit corrective actions The following are metrics examples that fall into the progress category:

- Cost performance index and variance

- Schedule performance index and variance

- Earned value

- Risk assessment tracking

- Manpower (planned versus actual)

- Deliveries

1.10.4 Product Metrics

Product metrics are measures of a program's technical maturity and are tied to the key performance parameters of a product. For developmental programs, these measures are found in the Operational Requirements Document (ORD) as objectives and thresholds and in the Test and Evaluation Master Plan (TEMP) as critical technical parameters Each performance parameter has an associated cost schedule, and risk impact. Metrics of this type indicate to teams whether or not the desired technical performance is achievable given the constraints of the program. To ensure a degree of commonality in reporting metric data to higher-level teams, the program team should determine the objectives that each sub-tier team is to accomplish, the frequency and level of detail of their reporting, and the allowed variation for each product metric. Examples of product metrics include --

- Operational availability

- Weight budget

- Mean time between failures (MTBF)

- Speed

- Range

- Payload

- Product unit cost

- Power consumption

1.10.5 Process Metrics

Process metrics assess the quality and productivity of a program's processes. In order to improve a process, it must be understood and measured. Data is collected at specific checkpoints in the process flow and then analyzed. The analysis of the data should be able to predict quality at later stages in the process;

Process metrics are a concern not only of the Integrated Product and Process Development (IPPD) stakeholders or IPTs measuring them, but also of the functional organizations (such as budgeting, contracting, or testing) that own the processes being measured. Cooperation is essential to ensure that the best metrics are used or developed.

Process metrics usually compare current/predicted performance versus performance objectives. A standard of performance is set using historical data or expected levels of performance. The process is then measured to see whether the objective is being met. If the objective is not met, analysis should determine why. If the objective is missed, it might suggest that the objective was not properly set. In either case, the process should be examined for ways to improve process performance and thereby establish a new objective. Statistical Process Control (SPC) is a good method to use for monitoring, controlling, and improving processes. Examples of process metrics are --

- Number and cost of requirements changes

- Number and cost of engineering change proposals

- Number and cost of test failures

- Cycle Time

- Defect rates

1.10.6 Metric Development Process

Choosing or creating metrics is not a random process. Developing a measurement system requires an in-depth understanding of customer and project requirements. Program processes and process outputs must be identified. From there, process output thresholds must be determined and the appropriate measures or performance indicators developed. The following nine-step process is not the definitive methodology for metric development, but it does provide guidance in what to consider when creating or choosing a metric, specifically for process metrics.

1. *Identify the purpose of the metric*. Is this metric intended to provide data only to the team creating it or will it be reported to higher level teams? What type of metric is needed -- programmatic/management control, technical performance measure, or process?

2. *Define what is to be measured*. Identify what it is that needs to be measured to satisfy the purpose (see step 1). If the process that is to be measured is not clearly understood in terms of cause-and-effect relationships, then the measurement will consist of a trial-and-

error determination of seemingly related factors that may or may not have a bearing on the outcome.

3. ***Identify and examine existing metrics.*** Once the cause-and-effect relationships have been identified, existing metrics from this or other programs should be examined to determine if any of them satisfies the requirement. It makes good sense to use a proven metric when the process previously measured matches or parallels the process under consideration.

4. ***Generate new metrics if existing metrics are inadequate.*** When generating a new metric, pay attention to what is needed as an output of the process to be measured and how that output contributes to the end product. With metrics, the focus is on a process' contribution to these final outputs. Teams should be interested in those measures that drive the final outcome and are key to making process improvements.

5. ***Rate the metric against the attributes of a good metric*** The metric should satisfy all of the criteria listed. If it does not, return to step 2 and correct the deficiencies.

6. ***Select the appropriate measurement tools.*** Keep in mind that the metric data should be economical to gather. This includes the hours spent gathering the data, processing it and the time required to display it. Automated data gathering is preferred, but many collection processes do not lend themselves to automation. Once the data is gathered, it often requires analysis or processing to be useful. There are many means of analyzing and displaying the data, such as process variance charts and control charts for process data. In some cases, a specialist may be needed to analyze and present the data.

7. ***Baseline the metric.*** This will serve as a reference point to begin acquiring data and measuring any changes.

8. ***Collect and analyze metric data over time.*** Aggregate metric data over time and examine trends. Special and/or common causes of effects on the data should be investigated. Compare the data with the baseline to ascertain improvement, decline, or no change. Utilize SPC when and as appropriate.

9. ***Initiate process improvement activities.*** Initiate iterative process improvement activities with key process owner involvement. Once the process has been changed, the data must be closely watched for trend improvement. If degradation is noticed, the reason for it must be identified and corrected. The process should not be changed until data trends have been clearly established, unless a change is required to correct a previous change that resulted in a decline in performance.

1.10.7 General Guidelines for Team Metrics

Metrics for IPT performance generally follow the guidelines below.

1. Metrics should measure only what is important, and output/outcome metrics are preferable to input/activity metrics. Metrics should measure the goals of the IPT as stated in the charter for the following reasons:

    • All activities of the IPT should be centered on meeting the chartered goals. Measurements of any other items are distractions. When the metrics apply directly to these goals, products of these goals -- which the program manager might be tempted to track independently -- will be tracked indirectly with the primary metrics.

    • Reporting and documentation of extraneous metrics consume too much valuable time that should be devoted to accomplishing the chartered goals.

    • Metrics should roll up through the IPT hierarchy. Metrics not directly related to primary goals cannot easily track upwards.

2. Metrics should be measurable in real time. The purpose of metrics is to indicate the current performance of the IPT. Metrics of a historical nature are just that -- a measure of what has happened, a summary, and not necessarily an indicator of future performance. Ideal measurements should also be easy to make. Difficult measurements take time and, thus, may not be done "in real time."

3. Metrics should be "on display." Metrics should be visible to those whose work is being tracked. This can give individual team members a better understanding of the goals of the team and what actions are needed to better meet those goals.

4. Metrics should be updated. Because acquisition programs change as time progresses and, thus, the goals of the IPTs change over time, metrics also need to be frequently reevaluated for current applicability.

2.0 SUSTAINING OPERATIONS

2.1 Introduction

Operational objectives are the execution of a support program that meets operational support performance requirements and sustainment of systems in the most cost-effective manner for the life cycle of the system. Three principal objectives with regard to A_O are:

1. Operate the system to achieve the design A_O;
2. Monitor the program A_O to identify deviations from plans and determine the degree of deviation; and,
3. Identify corrective actions to maintain the required A_O in deployed systems.

Mean Logistics Delay Time (MLDT), specifically MSRT, frequently has the single greatest impact on system A_O. While MTBF and MTTR are usually measured in minutes or hours; MLDT is often measured in days, weeks or months, and occasionally, years. There are several forces acting on system A_O and cost, which warrant management attention, such as:

- Configuration management in terms of repair part allowancing and stock control,
- Material availability after system modification/upgrade,
- Organic or contract supported physical distribution capability,
- Material obsolescence as systems age beyond traditional commercial lifespan,
- Material availability as Original Equipment Manufacturers (OEMs) consolidate or cease to operate,
- Material availability outside of traditional supply lanes,
- Physics of failure cause reductions in systems reliability,
- Systems aging factors cause reductions in systems reliability,
- Maintenance induced failures cause reductions in systems reliability, and
- Environmental conditions contribute to the above three.

2.2 A_O / Cost Study Objectives

The following strategies for improving A_O performance and/or cost savings apply:

- Validation of the A_O and cost estimates with actual fleet feedback data.
- Identify specific sub-systems and components that are driving A_O and cost problems.
- Provide decision support analysis and recommendations for system improvements.
- Confirm the achievement of A_O during early fielding. Measure the achieved A_O.
- Manage changes or modifications in design, configuration or support resources that impact the achievement of the A_O threshold.
- Identify resources to improve A_O if the A_O threshold is not being achieved.
- Identify post-production support issues.

2.3 Studies and Analyses

Methodologies for improving chances of success include:

- Ensure that the fielded system continues to provide the required A_O and cost characteristics.
- Ensure that the configuration, installation and Fleet operation of the system is consistent with the product specifications and use study from which the system was developed.
- Perform A_O and cost studies and analysis to support recommendations for system upgrades, modernization, technology insertion and other engineering modifications.

2.4 Monitoring Achieved A_O from Fleet Reporting

The process includes the study of the interdependent impacts of shortfalls in reliability, maintainability, or supportability upon each other, the A_O of the system sub-units, and ultimately, the material readiness of the system. The system can be displayed in a matrix with its major components on one axis and the components of A_O and their sub-elements on the second axis.

The program team should continue to monitor key indicators including those critical path items and resource requirements that vary from the required levels of performance. The program team assesses the impact of a variance in one component upon other components of A_O. The dependent relationships are important to the program team not only during production, but become critical to analysis of problems in the deployed operational system.

2.5 Assessing the Impact of Deviations, Changes and Modifications

During operations / sustainment, the need for modifications and/or engineering changes becomes more probable. These will naturally change the allocation of the Ao driving parameters (reliability, maintainability and supportability) among the components of the system.

2.6 Execute the Plan to Sustain A_O

The final test of whether the program team has maintained the system design A_O through production is the A_O achieved by the system when it is deployed in the operational environment. In order to achieve this, the program team has to establish the following:

- Performance of a continuing and consistent reporting system to monitor system performance.
- Compile those reports and determine how the Program Manager can monitor the system performance.
- Manage variances from established thresholds.

Inevitably, the parts required to restore operations will be for items not in stock. Throughout the operational cycle of the system, the program team's first and most critical task is to define the problem when the system fails to meet performance parameters, assess the impacts and decide when, and what, corrective actions are advisable.

2.7 Documentation Reports and Records

Several reports should be generated during the operational phase to address potential or actual problems, alternatives, and recommendations including problem identification and alternative analysis reports and decision support trade-study reports.

2.8 Follow-On Tracking

The primary element of documentation for this phase is the plan for the follow-on tracking, monitoring and reporting system for A_o and the components of A_o in the operational environment. The following key action steps are required, on a continuing basis, to execute the production, deployment, and follow-on support aspects:

- Manage changes/modifications in design, configuration or support resources (including contractor logistics support) to maintain the A_o threshold.
- Assess the impact on system A_o due to changes in the configuration.
- Identify resources to improve A_o, if the A_o threshold is not achieved.

2,9 Concept of Availability Contracting

The concept of Availability Contracting, a process by which defence contractors are paid according to the amount of time that an asset is available, is by no means a new idea for Western defence departments. Availability Contracting is an evolution of the idea of Contractor Logistic Support (CLS), by which services which would previously have been within the writ of military forces are outsourced to private contractors

The idea itself derives from the American concept of Performance Based Logistics (PBL). The essential premise of PBL is that contractors are paid accordance to measurable performance targets, for example the speed of repairs or relative cost effectiveness. These types of criteria are intended to ensure that contractors perform their tasks at least to the level of service that would have been provided by the equivalent military body.

2.10 Performance-Based Logistics (PBL)

PBL is a strategy for cost-effective system and supply product support. PBL can extend beyond traditional methods to integrate real-time 'lifecycle event data'. This has allowed information to be made available at different points in the lifecycle of an individual product or product stream.

PBL is the purchase of support as an integrated, affordable, performance package designed to optimize system readiness and meet performance goals for a weapons system through long-term support arrangements with clear lines of authority and responsibility. Simply put, performance based strategies buy outcomes, not products or services.

"Program Managers" shall develop and implement performance-based logistics strategies that optimize total system availability while minimizing cost and logistics footprint." "One of the most critical elements of a PBL strategy is the tailoring of metrics to the operational role of the system, and ensuring synchronization of the metrics with the scope of responsibility of the support provider."

2.11 Effective Application of Life Cycle Cost Management

Effective application of life cycle cost management accomplishes the following:

(a)   Provides basis for review and logical establishment of economically feasible performance requirements.

(b)   Provides design guidance to reduce ownership costs of achieving specific performance objectives.

(c)   Promotes innovations of lower ownership cost designs.

(d)   Provides a common basis for comparing a wide spectrum of design objectives.

2.11.1 Managing to Achieve Life Cycle Cost Objectives.

Life cycle cost efforts potentially add two new areas of activity to those previously assigned to a program office, i.e., life cycle cost estimating and life cycle cost management. The program manager must ensure that the required applicable actions are taken in these two areas. With respect to the primary area of life cycle costing activities addressed in this guide, i.e, designing the equipment to reduce costs, the manager's role is primarily to ensure that usual development and production functions are done with significantly increased emphasis on the consideration of life cycle cost implications of design and other decisions. This often includes the estimation and assessment of the life cycle cost differences among design and program alternatives.

Historically, performance objectives have been the first and most important criteria against which military equipment and in turn program personnel were assessed. Therefore, emphasis on meeting performance objectives often diverted design attention to this area, and away from factors which could reduce ownership costs. As a result, proper design attention was seldom given to reducing life cycle costs until after many design decisions, needed to meet performance objectives, had been made. Unfortunately, many of these performance-oriented design decisions significantly increased life cycle costs. Therefore, if a program manager is going to assure that life cycle costs, especially ownership costs, are adequately considered in design, he must take action from the outset of design activities to ensure that the designers are directed and motivated to look well beyond performance objectives in evolving the design.

One problem that may arise and cause difficulties is that a manager may find himself saddled with an impossible set of performance, cost and schedule objectives. This may occur as the result of new information provided by detail design studies or by development test results which indicate unanticipated problems. It may also occur as the result of planning studies which had to be based on preliminary or inadequate data, or unwarranted optimism. Whatever the cause, it is critical that a manager use the latest data available to continually keep the program proceeding toward a well balanced set of cost, schedule and performance objectives. This may require significant modification of earlier objectives based on later information. This is critical from a life cycle cost standpoint. Historically, when programs have continued to pursue impossible or nearly impossible design, performance or schedule objectives, they have generally resulted in systems and equipment being deployed with low reliability, high support costs and less than predicted performance. This, in turn, has resulted in high ownership costs and significant degradations in planned cost effectiveness.

During the design of systems and equipments, considerable testing is usually done to provide design decision guidance. Most of this testing is both time consuming and expensive. The result in the past has often been that test activities were curtailed when continued testing would have been a more appropriate course of action. Life cycle costs are closely related to the reliability of equipment as demonstrated in the field. Therefore, both realistic and adequate reliability tests are essential to develop and demonstrate equipment with satisfactory and known reliability characteristics prior to production commitment decisions. To achieve life cycle costing objectives, managers will have to make many difficult decisions concerning the conditions and duration of test programs, and what actions to take based on test results.

In addition, appropriate design studies and tests should be conducted to assure that the causes of premature wear, fatigue and corrosion-induced failures are eliminated from the design. A large fraction of current ownership costs is associated with non-failure maintenance actions such as inspection, adjustment and cleaning. Design studies to minimize these maintenance actions are also required.

An important point to remember is that the development of new equipment, which will be reliable and easy to maintain, usually involves significant design and development work solely to achieve the required reliability and maintainability objectives. Work up to the point of demonstrating adequate performance with some sort of prototype device, may be less than half the total effort required to develop and demonstrate a lower life cycle cost design. However, the overall design task will be accomplished most efficiently by considering life cycle cost related objectives from the outset of design, rather than waiting until after performance objectives have been demonstrated.

It is important that every manager clearly understand that program life cycle costing activities cannot be delegated to one or a small number of experts. Specialists may be required for some tasks, such as developing a life cycle cost model. However, if life cycle costing is to be successfully applied to the overall program throughout the acquisition process, all program and contractor personnel who make decisions affecting life cycle cost, must consider life cycle cost in arriving at decisions in their areas of responsibility. This includes personnel involved in planning, engineering, procurement, resource and decision analysis, and management.

2.11.2 Life Cycle Cost Analysis.

Feasible and efficient life cycle cost analysis methods will vary widely depending on the phase of the program, the nature of the equipment, the availability of data, the nature of decisions under study, and many other factors. Many life cycle cost design trade studies will require at least the adaptation of existing methods. Life cycle cost design trade studies for new products will often require development of new analysis methods or the adaptation of existing models reflecting the unique characteristics of the equipment.

The primary objective of life cycle cost design trade studies is to provide visibility with respect to the life cycle cost implications of various design and performance alternatives. Life cycle cost analysis of design alternatives requires knowledge of both design and cost analysis. Since engineers or design specialists have the primary role in evolving the final design, they can most effectively take the lead in performing beneficial life cycle cost design trade studies.

3.0 OPTIMIZATION

Optimization is an important tool in making decisions and in analyzing physical systems. In mathematical terms, an optimization problem is the problem of finding the best solution from among the set of all feasible solutions.

The optimization allows design and operation of systems or processes to make them as good as possible in some defined sense. The approaches to optimizing systems are varied and depend on the type of system involved, but the goal of all optimization procedures is to obtain the best results possible (again, in some defined sense) subject to restrictions or constraints that are imposed. While a system may be optimized by treating the system itself, by adjusting various parameters of the process in an effort to obtain better results, it generally is more economical to develop a model of the process and to analyze performance changes that result from adjustments in the model. In many applications, the process to be optimized can be formulated as a mathematical model; with the advent of high-speed computers, very large and complex systems can be modeled, and optimization can yield substantially improved benefits.

3.1 Designing an Optimal System

System design is one of the important applications of optimal decision-making problems. The fundamental goal of a system design is to build the system such that it performs its functions successfully. The inability of the system to perform its functions is called a system failure. Several factors related to system design as well as external events influence the system functionality. In most cases, the effects of these factors are random, which means that they cannot be determined precisely but can only be explained through probability distributions. Therefore, the failure event or the time to system failure is a random variable.

The engineering discipline that deals with the successful and unsuccessful or failure operations of systems is known as reliability engineering. Reliability is a critical system characteristic and is defined as the "probability that the system performs its intended or specified functions successfully over a specified period of time under the specified environment." One of the goals of reliability engineering is to carefully design and analyze a system design to build the highest reliability into the system within the limits of economical and physical constraints. Some important principles for enhancing system reliability follow.

- Keep the system as simple as possible while still meeting all performance requirements. This can be achieved by minimizing the number of components in series and their interactions.

- Increase the reliability of the components in the system. This can be achieved by reducing the variations in the components' strength and applied load through better quality control and monitoring of operational environment, increasing the strength of the components by substituting better materials, and reducing the applied load. Alternatively, increasing component reliability can be achieved by using large safety factors or management programs for product improvement.

- Use burn-in procedures for components that have high infant mortality to eliminate early failures in the field.

- Use redundancy, or spares, for less reliable components. This can be achieved by adding spares in the parallel or standby redundancy.

- Develop a fault-tolerant design such that the system can continue its functions even in the presence of some failures. This can be achieved using sparing redundancy, fault-masking, and failover capabilities.

Additionally, if the system or its components are repairable, the availability of the system should be considered as a system performance index. Availability of the system is "the probability that the system is operational at a specified time." In a long run, the system availability estimate reaches an asymptotic value called steady-state availability. In most cases, reliability engineers focus their attention on improving the steady-state availability of the system, which can be accomplished by reducing the downtime. Some important principles for enhancing the reliability of a repairable system follow.

- Use design methods that increase the reliability of the system.

- Decrease the downtime by reducing delays in performing repairs. This can achieved by keeping additional spares onsite, providing better training for repair personnel, using better diagnosis procedures, and increasing the number of repair personnel.

- Perform preventive maintenance such that components are replaced by new ones whenever they fail, or at some regular time intervals or age, whichever comes first.

- Perform condition-based maintenance such that downtime related to either preventive or corrective maintenance is minimal.

- Use better arrangements of exchangeable components.

Implementation of the above principles to improve system performance such as reliability or availability typically consumes some resources, which may be limited. Resource limitations may include available budget, space to keep components, and weight limitations. In such cases, the objective should be obtaining the maximum system performance within the utilization of the available resources. However, in some cases, achieving high performance may not lead to the maximum overall profit or minimum overall cost. In such cases, the system design should be optimized to achieve the most cost-effective solution that strikes a balance between the system, the cost of system failure, and the cost of efforts for reducing system failures.

3.2 Where Optimization Used

Optimization is applied in virtually all areas of human endeavor, including engineering system design, optical system design, economics, power systems, water and land use, transportation systems, scheduling systems, resource allocation, personnel planning, portfolio selection, mining operations, blending of raw materials, structural design, and control systems. Optimizers or decision makers use optimization in the design of systems and processes, in the production of products, and in the operation of systems.

3.3 Constructing a Model

The first step in the optimization process is constructing an appropriate model; modeling is the process of identifying and expressing in mathematical terms the objective, the variables, and the constraints of the problem.

- An *objective* is a quantitative measure of the performance of the system that we want to minimize or maximize. In manufacturing, we may want to maximize the profits or minimize the cost of production, whereas in fitting experimental data to a model, we may want to minimize the total deviation of the observed data from the predicted data.
- The *variables* or the *unknowns* are the components of the system for which we want to find values. In manufacturing, the variables may be the amount of each resource consumed or the time spent on each activity, whereas in data fitting, the variables would be the parameters of the model.
- The *constraints* are the functions that describe the relationships among the variables and that define the allowable values for the variables. In manufacturing, the amount of a resource consumed cannot exceed the available amount.

3.3.1 Continuous versus Discrete Optimization

Some models only make sense if the variables take on values from a discrete set, often a subset of integers, whereas other models contain variables that can take on any real value. Models with discrete variables are *discrete optimization* [*https://neos-guide.org/content/discrete-optimization*] problems; models with continuous variables are *continuous optimization* [*https://neos-guide.org/content/continuous-optimization*] problems. Continuous optimization problems tend to be easier to solve than discrete optimization problems; the smoothness of the functions means that the objective function and constraint function values at a point can be used to deduce information about points in a neighborhood of . However, improvements in algorithms coupled with advancements in computing technology have dramatically increased the size and complexity of discrete optimization problems that can be solved efficiently. Continuous optimization algorithms are important in discrete optimization because many discrete optimization algorithms generate a sequence of continuous subproblems

3.3.2 Deterministic versus Stochastic Optimization

In deterministic optimization, it is assumed that the data for the given problem are known accurately. However, for many actual problems, the data cannot be known accurately for a variety of reasons. The first reason is due to simple measurement error. The second and more fundamental reason is that some data represent information about the future (e. g., product demand or price for a future time period) and simply cannot be known with certainty. In *optimization under uncertainty*, [*https://neos-guide.org/content/optimization-under-uncertainty*] or *stochastic optimization*, the uncertainty is incorporated into the model. Robust optimization techniques can be used when the parameters are known only within certain bounds; the goal is to find a solution that is feasible for all data and optimal in some sense. Stochastic programming [*https://neos-guide.org/content/stochastic-programming*] models take advantage of the fact that probability distributions governing the data are known or can be estimated; the goal is to find some policy that is feasible for all (or almost all) the possible data instances and optimizes the expected performance of the model.

3.3.3 Determining the Problem Type

The second step in the optimization process is determining in which category of optimization your model belongs. The URL *https://neos-guide.org/optimization-tree* provides some guidance to help classify your optimization model; for the various optimization problem types, there is a linked page with some basic information, links to algorithms and software, and online and print resources.

3.3.4 Selecting Software

The third step in the optimization process is selecting software appropriate for the type of optimization problem that you are solving. Optimization software comes in two related but very different kinds of packages:

- *Solver software* is concerned with finding a solution to a specific instance of an optimization model. The solver takes an instance of a model as input, applies one or more solution methods, and returns the results.
- *Modeling software* is designed to help people formulate optimization models and analyze their solutions. A modeling system takes as input a description of an optimization problem in a symbolic form and allows the solution output to be viewed in similar terms; conversion to the forms required by the algorithm(s) is done internally. Modeling systems vary in the extent to which they support importing data, invoking solvers, processing results, and integrating with larger applications. Modeling systems are typically built around *a modeling language* for representing the problem in symbolic form. The modeling language may be specific to the system or adapted from an existing programming or scripting language.

Most modeling systems support a variety of solvers, while the more popular solvers can be used with many different modeling systems. Because packages of the two kinds are often bundled for convenience of marketing or operation, the distinction between them is sometimes obscured, but it is important to keep in mind when attempting to sort through the many alternatives available.

3.3.5 Commercial vs. Open Source Solvers

Commercial solvers are developed with considerable effort and, while usually more robust and reliable, they often are quite expensive. Some commercial systems are available for free under reasonable conditions for educational purposes and academic research. Many offer free size-limited student (or demo) versions for experimentation with small problem instances. Open source solvers make their source code freely available under one of the standard open source licenses; many of these are available through the COIN-OR repository (www.coin-org.org). Many of the open-source solvers are also available as precompiled binaries for the more popular platforms.

https://neos-guide.org/content/optimization-introduction
https://neos-guide.org/Optimization-Guide

3.3.6 Models Classification in Optimization

System models used in optimization are classified in various ways, such as linear versus nonlinear, static versus dynamic, deterministic versus stochastic, or time-invariant versus time-varying. In forming a model for use with optimization, all of the important aspects of the problem should be included, so that they will be taken into account in the solution. The model can improve visualization of many interconnected aspects of the problem that cannot be grasped on the basis of the individual parts alone. A given system can have many different models that differ in detail and complexity. Certain models (for example, linear programming models) lend themselves to rapid and well-developed solution algorithms, whereas other models may not. When choosing between equally valid models, therefore, those that are cast in standard optimization forms are to be preferred. *See*

The model of a system must account for constraints that are imposed on the system. Constraints restrict the values that can be assumed by variables of a system. Constraints often are classified as being either equality or inequality constraints. The types of constraints involved in any given problem are determined by the physical nature of the problem and by the level of complexity used in forming the mathematical model.

Constraints that must be satisfied are called rigid constraints. Physical variables often are restricted to be nonnegative; for example, the amount of a given material used in a system is required to be greater than or equal to zero. Rigid constraints also may be imposed by government regulations or by customer-mandated requirements. Such constraints may be viewed as absolute goals.

In contrast to rigid constraints, soft constraints are those constraints that are negotiable to some degree. These constraints can be viewed as goals that are associated with target values. The amount that the goal deviates from its target value could be considered in evaluating trade-offs between alternative solutions to the given problem.

When constraints have been established, it is important to determine if there are any solutions to the problem that simultaneously satisfy all of the constraints. Any such solution is called a feasible solution, or a feasible point in the case of algebraic problems. The set of all feasible points constitutes the feasible region.

3.4 Optimization of Availability with Prognostics

The Operational Availability optimization requires minimizing the sum of Preventive and Corrective Maintenance costs. Such a minimum per unit time, at a certain time, it can be expressed as a function between the Mission Time value, the ratio between Corrective and Preventive Maintenance costs and of the System Reliability. The Mission Time's lifespan is a known data and is a part of mission profile figure. Preventive Maintenance cost includes spare parts costs and labor costs. Corrective Maintenance includes repair task costs such as spare parts costs, labor costs and eventual supplementary costs due to adjunctive failures and includes the cost of lost system operation; those quantities costs are generally known as economic analyses.

Instead, for time-limited items subject to material degradations, a precise and accurate reliability assessment in time is a matter of difficult concern because it is necessary to introduce the degradation material laws and the uncertainties on the rated value of the parameters used as input.

3.4.1 Prognostics Explained

The prognostic is the branch of the science that study the evolution of the process and it is quantified by the reliability prediction performed, based on design specifications. In the engineering fields, mechanical, electrical and even in electronic, the philosophical approach until few time ago , forecasts the reliability prediction on the base of statistical data obtained by field and laboratory tests.

Only recently, the process simulation is strongly entry in the diagnostic and in prognostic also. In engineering, it is in fact, it is very much easier to express in mathematical language the models concerning the mode operations of each component, of the whole complex system and also to link the failure modes of components to failure mode of the system. This link allows calculating with sufficient precision and accuracy the EOL (Expected operation life) and the RUL (remained useful life).

3.4.2 Prognostics and Health Management

Prognostics and Health Management (PHM) is an approach that is used to evaluate the reliability of a system in its actual life-cycle conditions, to determine the initiation of failure, and to mitigate the system risks. Prognostics of a system can yield an advance warning of impending failure in the system and thereby help taking appropriate corrective actions. It helps in preventing catastrophic failures and reduces unscheduled maintenance expenses. Prognostics have become the preferred approach to achieve efficient system-level maintenance and reduce the life cycle cost of the system The Unites States Department of Defense' 5000.2 policy document on defense acquisition, states that program managers should utilize diagnostics and prognostics to optimize the operational readiness of defense-related systems

The approaches adopted for conducting prognostics for a product are: (1) use of canaries to provide advance warning of failure, (2) monitoring the precursors to impending failure, and (3) modeling of life cycle environment stress to compute accumulated damage [4]. Monitoring of precursors and modeling of life cycle environment involves sensing parameters related to the product and environment and using predictive models to estimate the damage to the product.

PHM exercises necessary management actions. Prognostics utilizes in-situ monitoring and analysis to assess system degradation, and determine the Remaining Useful Life (RUL) of an asset. The RUL of an asset is defined as the length of time from the present time to the end of useful life. The need for RUL estimation is obvious because it relates to a frequently asked question in industry, which is how long a monitored asset can survive based on the available information. Based on the RUL estimation, appropriate actions can be planned. Especially for critical equipment, such as aircraft engines, or inertial navigation platforms used in aerospace and weapon systems, determining if and when to take equipment out of service is important from both a cost-effective point of view and a safety point of view.

It is critically important to assess the RUL of an asset while it is in use, as this information impacts the planning of maintenance activities, the supply chain, replenishment of the inventory system, operational performance, and profitability of the owner of an asset
.
To learn and apply PHM methodology, recommend joining the CALCE Prognostic and Health Management Group http://www.prognostics.umd.edu/

3.5 If no Solution for an Optimization

If no feasible solution exists for a given optimization case, the decision maker may relax some of the soft constraints in an attempt to create one or more feasible solutions; a class of approaches to optimization under the general heading of goal programming may be employed to relax soft constraints in a systematic way to minimize some measure of maximum deviations from goals.

A key step in the formulation of any optimization problem is the assignment of performance measures (also called performance indices, cost functions, return functions, criterion functions, and performance objectives) that are to be optimized. The success of any optimization result is critically dependent on the selection of meaningful performance measures. In many cases, the actual computational solution approach is secondary. Ways in which multiple performance measures can be incorporated in the optimization process are varied

4.0 CRITICAL TECHNICAL PROCESSES RELATED TO AVAILABILITY

The following contains fundamental technical processes and the associated "Best Practices" and "Watch-Out-Fors" which have great influence on technical risk. These practices, though by no means comprehensive, do focus on key technical risk areas. Use of proven best practices to achieve product success leads to a more organized approach to accomplish these activities and places more management significance on them.

The "Best Practices" and "Watch-Out-Fors" associated with critical industrial technical processes should be used as a starting point in developing a baseline of program specific contractor processes. The best practices associated with these critical processes can also serve as benchmarks with which to compare your program's baseline processes and results achieved versus desired goals. The following examples of critical processes for the Design, Test, and Production phases of a product's development are presented in this chapter

4.1 Design Reference Mission Profile

A Design Reference Mission Profile (DRMP) is a hypothetical profile consisting of time-phased functional and environmental profiles derived from multiple or variable missions and the total envelope of environments to which the system will be exposed. The DRMP becomes the basis for system and subsystem design and test requirements.

Best Practice

- Mission Profiles cover all system environments during its life cycle including operational, storage, handling, transportation, training, maintenance, and production

- Mission Profiles are defined in terms of time (duration and sequence), level of severity, and frequency of cycles

- Mission and System Profiles are detailed by the Government and contractor respectively, based on natural and induced environments (e.g., temperature, vibration, electromagnetic impulse, shock, and electrical transients)

- Profiles are the foundation for design and test requirements from system level to piece parts, including Commercial-Off-The-Shelf/Non-Developmental Items (COTS/NDIs)

Watch Out For

- DRMP environmental profiles that appear to be simply extracted from MIL-HDBK 810, "Environmental Test Methods and Engineering Guidelines," 31 July 1995

- Mission Profiles based on average natural environmental conditions rather than the more extreme conditions that may more accurately reflect operational requirements in the place/at the time of use, such as indicated by MIL-HDBK-310 "Global Climatic Data for Developing Military Products," 23 June 1997 and the National Climatic Data Center

4.2 Trade Studies

Trade Study are iterative series of studies performed to evaluate and validate concepts representing new technologies or processes, design alternatives, design simplification, ease of factory and field test, and compatibility with production processes. Trade studies culminate in a design that best balances need against what is realistically achievable and affordable.

Best Practice

- Trade studies are performed to evaluate alternatives and associated risks

- Trade studies consider producibility, supportability, reliability, cost and schedule as well as performance

- Trade studies are conducted using principles of modeling and simulation, experimental design and optimization theory

- Trade studies include sensitivity analyses of key performance and life cycle cost parameters

- Trade study alternatives are documented and formally included in design review documentation to ensure downstream traceability to design characteristics

- Trade studies are traceable to the DRMP and associated design requirements

- Quality Function Deployment techniques are used to identify key requirements when performing trade-offs

Watch Out For

- Use of new technologies without conducting trade-studies to identify risks

- Trade studies that do not include participation by appropriate engineering disciplines

- Product reliability, quality and supportability traded for cost, schedule and functional performance gains

4.3 Design Analyses

Design Analyses are performed to examine design parameters and their interaction with the environment. Included are risk-oriented analyses such as stress, worst case, thermal, structural, sneak circuit, and Failure Modes, Effects and Criticality Analysis (FMECA), which, if conducted properly, will ensure that reliable, low risk, mature designs are released.

Best Practice

- Validate new analysis modeling tools prior to use

- Conduct logic analysis on 100% of Integrated Circuits (ICs)

- Analyze 100% of IC outputs for ability to drive maximum expected load at rated speed and voltage levels

- Use Table 4.3-1 below to determine which design analyses should be performed

Watch Out For

- Analyses performed by inexperienced analysts

- Analyses performed using unproven software programs

Table 4.3-1. Objectives of Selected Design Analyses

Analyses	Objectives
• Reliability Prediction	• To evaluate alternative designs, assist in determining whether or not requirements can be achieved and for help in detecting over-stressed parts and/or critical areas
• Failure Modes, Effects and Criticality Analysis	• To identify design weaknesses by examining all failure modes using a bottom-up approach
• Worst Case Analysis	• To evaluate circuit tolerances based on simultaneous part variations
• Sneak Circuit Analysis	• To identify latent electrical circuit paths that cause wanted functions or inhibit wanted functions
• Fault Tree Analysis	• To identify effects of faults on system performance using a top-down approach
• Finite Element Analysis	• To assure material properties can withstand intended mechanical stresses in the intended environments
• Stress Analysis	• To determine or verify design integrity against conditional extremes or design behavior under various loads
• Thermal Stress Analysis (see Thermal Analysis)	• To determine or eliminate thermal overstress conditions; to verify compliance with derating criteria

4.4 Parts and Material Selection

The Parts and Material Selection utilizes a disciplined design process including adherence to firm derating criteria and the use of Qualified Manufacturers Lists (QML) to standardize parts selection.

Best Practice

- Use (QML) parts, particularly for applications requiring extended temperature ranges

- Electrical parameters of parts are characterized to requirements derived from the Design Reference Mission Profile to ensure that all selected parts are reliable for the proposed application.

- Derate all parts electrically and thermally

- A Preferred Parts List is established prior to detailed design

- Parts screening is tailored based on maturity

- Use highly integrated parts (e.g., Application Specific ICs (ASICs)) to reduce:

  - The number of individual discrete parts/chips
  - The number of interconnections
  - Size, power consumption, and cooling requirements, and
  - Failure rates

- Quality is measured by:

  - Certification by supplier

  - Compliance with JEDEC-EIA-623, "Procurement Quality of Solid State Components by Governments Contractors," July 1994 [https://standards.globalspec.com/std/330160/jedec-eia-623]

  - Verification to historical data base

  - Particle Impact Noise Detection for cavity devices

  - Destructive Physical Analysis for construction analyses

- Strategy for parts obsolescence and technology insertion is established

- Vendor selection criteria established for non-QML parts considering:

  - Qualification, characterization and periodic testing data
  - Reliability/quality defect rates
  - Demonstrated process controls and continuous improvement program
  - Vendor production volume and history

- Minimum acceptable defects for in-coming electronic piece parts:

  - Maximum of 100 defective parts per million

Watch Out For

- Development of highly integrated parts unique to one specific acquisition development program

- Use of non-QML parts whenever QML parts are available

- Highly integrated parts that are not treated as a system of discrete parts to which the parts program requirements also apply

- Use of parts in environments not specified by the Original Equipment Manufacturer

- Variance in operating characteristics of commercial RF and analog parts

- Use of any parts near, at, or above their rated values, especially plastic encapsulated devices which reach higher junction temperatures than ceramic devices due to higher resistance to heat conduction

- Device frequency derating based on maximum overall operating temperature vs frequency rating which varies at different operating temperatures

- The use of parts beyond specified operating ranges by upscreening or uprating

- Designs using part technologies whose remaining life cycle will not support production and postproduction uses

4.5 Testability

Testability refers to the design parameter which makes it relatively easy to identify and isolate faults in the system. In addition to detecting existing faults, the diagnostic capability of a testable system should also be able to detect incipient faults so that appropriate preventive action can be taken before the system fails. Testability can be considered to be a subset of maintainability, because fault detection and isolation are important drivers in the maintainability of a system. Testability also impacts system reliability because it affects "could not duplicate" and "re-test OK" malfunctions. Reducing this type of reported malfunction eliminates unnecessary maintenance actions and reduces the occurrence of repeat and recurring discrepancies.

4.5.1 Testability Metrics.

Measures of testability include time measurements, such as system downtime for test, labor hours spent in fault isolation, etc. Other measures more directly related to system design performance have also been developed. These include such metrics as fraction of faults detected, fraction of faults isolated, and false alarm rate. Application of testability metrics must be done with care. Acquisition reform requires that we specify the test and diagnostics system in performance terms, but we have to be certain to make the performance specification meaningful. Beware of percent goals. A poorly worded goal might be measurable, but, even though it looks good on paper, its implementation in the system may leave an operationally unacceptable test and diagnostics system. For example, a large aircraft test and diagnostics system could be monitoring several thousand data points continuously. A two-percent false alarm rate might sound great, but could result in almost continuous reporting of non-existent faults. This would drive the operators crazy, force huge amounts of unnecessary maintenance, create an astronomical "Re-test OK" rate, and mask any genuine faults in a flood of false alarms.

4.5.2 Accessibility.

When we refer to accessibility in the context of testability, we mean that the testing subsystem must be accessible to the technician under all operational environmental conditions, including the wear of personal protective equipment. Test points and ports for external test equipment should be readily available for use. Any built-in test equipment also needs to be readily accessible if it is to have any value. One method of enhancing the accessibility of built-in test equipment is to transmit the test results, encrypted as needed, from the airborne platform to a ground maintenance computer system so that the maintenance organization knows the equipment status and diagnosis before the air vehicle returns from its mission. This system provides accessibility to the test equipment while the air vehicle is still in flight. This allows the appropriate technician to meet the aircraft with all required repair parts and test equipment, greatly shortening turn time. Whatever method is used, the system design must allow the technician to easily hook up external test equipment if needed and to easily obtain the results of both external and built-in test equipment.

4.5.3 Visibility.

Visibility is the subset of accessibility that is most applicable to obtaining the results of the testing performed by built-in or external test equipment. Visibility implies that the results are easy to observe, even when darkness, other environmental conditions, or personal protective equipment interfere with normal vision. Visibility also implies that the results should be displayed in such a way that they are legible and unambiguous. The technician should be able to look at the display and immediately know the condition of the system and which part, down to the component or circuit card level, needs to be replaced to restore full system operation. The test equipment should display the information in such a clear and unambiguous format that a minimally trained technician can easily service the system without having to refer to paper technical orders or without consulting more skilled specialists.

4.5.4 Built-in Test Equipment (BITE).

The use of BITE is one way to increase system mobility by reducing the amount of support equipment required to keep the system operational. BITE also can improve diagnostic capability since it can monitor and record system performance under operational conditions. This capability should help eliminate could not duplicate (CND) and re-test OK (RTOK) discrepancies, since the system itself will report the exact problem without having to rely on the operator's memory. Also, as mentioned above, BITE may be combined with remote reporting of aircraft status to reduce aircraft turn times.

4.5.5 Automatic Test Equipment (ATE).

Automation is one way to take the variability associated with the human element out of a process. The use of ATE, whether built-in or external, eliminates some human judgment from the testing process. It does this by automatically detecting system faults, reporting status, and isolating the fault to the component level. Automating the fault detection and isolation process also impacts the skill level required by technicians servicing the system, and this, in turn, eliminates some training which would otherwise be needed. If the system itself has the artificial intelligence to perform its own test and diagnostic functions, the expertise required of the humans servicing the system is decreased. If the ATE is either built-in to the system, or if it can be used by relatively unskilled, organizational level technicians in the on-equipment environment, ATE becomes an enabling technology for the two-level maintenance concept. We should try to consider these aspects from the earliest stages of system development so that BITE and ATE can be designed in from the start. In this way, we can insure that the optimum use is made of ATE and BITE to achieve a testable system which conforms to current Air Force supportability policy concerns.

4.5.6 Self Calibration.

Ideally, the test equipment used to monitor system performance will not impose new maintenance requirements, such as periodic inspection and calibration. If the equipment needs periodic calibration, the time required for this maintenance decreases system availability. Also, calibration usually requires special test equipment and comparison standards which may need to be maintained in a laboratory environment. These requirements could have a serious adverse impact on weapon system mobility and supportability in the field. For this reason, built-in and external test equipment should employ self calibration wherever possible. This means that the equipment must contain its own reference standard or refer to some external, readily available standard, e.g. a time signal, to automatically calibrate and adjust its performance without outside intervention.

4.5.7 Diagnostics.

The diagnostic portion of a test system is that subset of the system that allows isolation of the fault or impending fault to the component that needs to be replaced. Diagnostics in this sense includes both fault isolation of a failed system and the prognostic ability to monitor system parameters to predict system failure and direct preventive maintenance action before the failure actually occurs. The ideal diagnostic system will always identify the correct component without the need for a highly trained specialist to interpret system symptoms. The diagnostic/prognostic subsystem is an ideal candidate for the application of artificial intelligence to eliminate the errors that can come from human fallibility.

4.6 Design for Testability

Designing for Testability assures that a product may be thoroughly tested with minimum effort, and that high confidence may be ascribed to test results. Testing ensures that a system has been properly manufactured and is ready for use, and that successful detection and isolation of a failure permits cost-effective repair.

Best Practice

- Perform testability analyses concurrently with design at all hardware and all maintenance levels

- Use Fault Tree Analysis, FMECA, and Dependency Modeling & Analysis to determine test point requirements and fault ambiguity group sizes

- Use standard maintenance busses to test equipment at all maintenance levels

- Use ASICs and other complex integrated circuits/chips with self-test capabilities

- Good testability design reflects the ability to:

  - Initialize the operating characteristics of a system by external means, e.g., disable an internal clock

  - Control internal functions of a system with external stimuli, e.g., break up feedback loops

  - Selectively access a system's internal partition and parts based on maintenance needs

- Evaluate Printed Wiring Board (PWB) testability using RAC publication Testability and Assessment Tool," 1991:

- Converts scored and weighted rating of factors, including accessible vs. inaccessible nodes, proper documentation, complexity, removable vs. non-removable components, and different logic types (34 factors in all), to a possible total score of 100

- The following testability scores illustrate this method

T-Scores for PWB Testability	
Acceptable Score	**PWB Testability**
81 to 100	Very easy
66 to 80	Easy
Questionable Score	**PWB Testability**
46 to 65	Some difficulty
31 to 45	Average difficulty
Unacceptable Score	**PWB Testability**
11 to 30	Hard
1 to 10	Very hard
0 or less	Impossible to test/troubleshoot w/out cost penalties

Watch Out For

- Incompatibility between operational time constraints and time required to perform Built In Test (BIT)

- COTS/NDI testability design that is incompatible with mission needs and program life-cycle maintenance philosophy

- Testability design that results in special-purpose test equipment

- Circuit card assemblies and modules with test points that aren't accessible

- Circuit functions that don't fit on a single board

- Reverse funneling of tests

- Testability requirements for production are defined after design release

4.7 Built In Test

Built-In-Test (BIT) provides "built in" monitoring and fault isolation capabilities as integral features to the system design. BIT can be supplemented with embedded "expert system" technology that incorporates diagnostic logic/strategy into the prime system.

Best Practice

- BIT is compatible with other Automatic Test Equipment (ATE)

- Use BIT software:

- For most flexible options (voting logic, sampling variations, filtering, etc.) to verify proper operation and identification of a failure or its cause

- To minimize BIT hardware

- To record BIT parameters

- Use multiplexing to simplify BIT circuitry

- Size the fault ambiguity group considering:

    - Mission requirements for reliability, repair time, down time, false alarm rate, etc.

    - Requirements for test equipment/manning at intermediate and depot maintenance levels

- Verify adequacy of the BIT circuit thresholds during development testing

- BIT should, as a minimum, provide:

    - 98% detection of all failures

    - Isolation to the lowest replaceable unit

    - Less than 0.1% false alarms

- Ratio of predicted to actual testability results 1:1

- Preliminary testability analysis -- completed before PDR

- Detailed testability analysis -- completed before CDR

Watch Out For

- High BIT effectiveness resulting in unacceptably high false alarm rates

- Inadequate time to perform BIT localization/diagnosis resulting in diminished BIT coverage and accuracy

- BIT design and analyses that fail to consider the effects of DRMP and worst-case variations of parameters, such as noise, part tolerance, and timing, especially as affected by age

- Inadequate BIT memory allocation

- Limitations to BIT coverage/effectiveness caused by:

    - Non-detectable parts (mechanical parts, redundant connector pins, decoupling capacitors, one-shot devices, etc.)

    - Power filtering circuits

- Use of special test equipment (e.g., signal generators) to simulate operational input circuit conditions

- Interface and/or compatibility problems between some equipment designs (e.g., digital vs analog)

- Unkeyed test connectors

- Test points without current limits

- Test points that are not protected against shorts to either adjacent test points or to ground

- Testing constraints that cause failures of one-shot devices, safety related circuits and physically restrained mechanical systems

- Methodology used to calculate BIT effectiveness

4.8 Design Reviews

A Design Review is a structured review process in which design analysis results, design margins and design maturity are evaluated to identify areas of risk, such as technology, design stresses, and producibility, prior to proceeding to the next phase of the development process.

Best Practice

- Formal procedures are established for Design Reviews

- Design Reviews are performed by independent and technically qualified personnel

- Entry and exit criteria are established

- Checklist and references are prepared

- Manufacturing, product assurance, logistics engineering, cost and other disciplines have equal authority to engineering in challenging design maturity

- Design Review requirements are flowed down to the subcontractors

- Subcontractors and customers participate in the design reviews

- Conduct design reviews as follows:

  - PDR: 20% of the design is complete
  - IDR: 50% of the design is complete
  - CDR: 95% of the design is complete

Watch Out For

- Reviews that are primarily programmatic in nature instead of technical

- Review schedules that are based on planned milestone dates

- Reviews held without review of analyses, assumptions, and processes

- Reviews held without review of trade-off studies, underlying data and risk assessments

- Reviews not formally documented and reported to management

- Reviews held by teams without adequate technical knowledge or representation of manufacturing, product assurance, supportability, etc.

4.9 Thermal Analysis

Thermal Analysis is one of the more critical analyses that is performed to eliminate thermal overstress conditions and to verify compliance with derating criteria. Thermal analyses are often supplemented with infrared scans, thermal paint, or the use of other measurement techniques to verify areas identified as critical.

Best Practice

- Determination and allocation of thermal loads and cooling requirements to lower-level equipment and parts are made based on the DRMP and the system self-generated heat

- Preliminary analyses are refined using actual power dissipation results as the thermal design matures

- The junction-to-case thermal resistance values of a device are used for the thermal analysis

- Thermal Survey (e.g., infrared scan) is conducted to verify the analysis

Watch Out For

- The use of device junction-to-ambient values for the thermal analysis, since this method is highly dependent on assumptions about coolant flow conditions

- A thermal analysis that does not take into account all modes (convection, conduction, radiation) and paths of heat transfer

4.10 Design Release

Design release is the point in the developmental stage of a product when creative design ceases and the product is released to production. Scheduling a design release is closely related to the status of other design activities such as design reviews, design for production, and configuration management.

Best Practice

- Design release process requires concurrent review by all technical disciplines

- Measurable key characteristics and parameters are identified on drawings, work instructions and process specifications

- Designs are released to production after:

    - Completion of all design reviews
    - Closeout of all corrective action items
    - Completion of all qualification testing

- A producible, supportable design is characterized by:

    - Stable design requirements

    - Completed assessment of design effects on current manufacturing processes, tooling and facilities

    - Completed producibility analysis

    - Completed rapid prototyping

- Completed analysis for compatibility with:

    - COTS/NDI interfaces

    - Subcontractor design interfaces

    - Form, Fit, and Function at all interfaces

- Design release practices, or equivalent, of the prime contractor are flowed down to the subcontractors

Watch Out For

- Design release based on manufacturing schedule

- Manufacturing drawings containing redlines

- Procurement for long lead items initiated with immature designs

- Drawings that are approved for release by engineering without review by all technical disciplines

4.11 Computer Aided Design / Computer Aided Manufacturing

Computer Aided Design/Computer Aided Manufacturing (CAD/CAM) introduces technical discipline throughout the design process to ensure success in complex development projects by integrating various design processes onto a common database. Included is the capability to perform special analyses, such as stress, vibration, thermal, noise, and weight, as well as to permit simulation modeling using finite element analysis and solids modeling. The outputs of this common database control manufacturing processes, tool design and design changes.

- Embed design rules in the CAD/CAM system

- Map CAD/CAM tools to the design and manufacturing processes

- Use compatible tools in an integrated CAD/CAM approach

- Use open architecture approach for software programs and data files

- Use new machine tools capable of being networked or upgraded to a network

- As a basis for procurement of new or upgraded CAD/CAM systems, sensitivity analyses are performed for various future scenarios (e.g., mainframe based versus Unix workstation-based, or NT based versus future cost to maintain and interconnect, 64 bit versus 32 bit math, links to ERP systems, etc.)

- 80% of design activity is computer based

- 100% of CAD drawings are CAM compatible

- Use common data exchange standards for 75% of processes

- All new machines networkable for CAM

Watch Out For

- CAD/CAM tools that operate in a stand-alone manner

- Failure to include total factory requirements and planned use for the CAD/CAM database

- Lack of a long-term growth plan to keep from being backed into a technological dead-end

- Proprietary Computer Numerically Controlled and Direct Numerically Controlled platforms and software architectures

- CAD/CAM systems which are non-standard for your industry, customers and suppliers

- Companies who will not be in business in several years

4.12 Design Limit Qualification Testing

Design Limit Qualification Testing is designed to ensure that system or subsystem designs meet performance requirements when exposed to environmental conditions expected at the extremes of the operating envelope, the "worst case" environments of the DRMP.

Best Practice

- Design limit/margin testing based on the DRMP, is integrated into the overall test plan, especially with engineering, reliability growth and life testing

- Design limit qualification tests are performed to ensure worst case specification requirements are met

- Highly Accelerated Life Tests (HALT) are performed to determine the design margins:

  - When operating at the expected worst case environments and usage conditions

  - To identify areas for corrective action

- Increased stress to failure conditions are included toward the end of Test, Analyze, and Fix (TAAF) testing to identify design margins

- Engineering development tests are performed beyond the design limits to measure the variance of the functional performance parameters under environmental extremes

- The failure mechanism of each failure, including stresses at the worst case specification limits, is understood

Watch Out For

- Design limit qualification testing environmental limits that are based on MIL-STDs and do not consider the DRMP

- In-service use of design limit qualification test units and other units that are stressed to a level resulting in inadequate remaining life

- Incompatibility of the COTS/NDIs qualification tests to the requirement

- Accelerated testing conditions which introduce failure modes not expected in normal use

4.13 Test, Analyze, and Fix

The Test, Analyze and Fix (TAAF) process is an iterative, closed loop reliability growth methodology. TAAF is accomplished primarily during engineering and manufacturing development. The process includes testing, analyzing test failures to determine cause of failure, redesigning to remove the cause, implementing the new design, and retesting to verify that the failure cause has been removed.

Best Practice

- Use of Duane or AMSAA Growth Models for the TAAF process

- Test facilities are capable of simulating all environmental extremes

- TAAF process starts at the lowest level of development and continues incrementally to higher assembly levels through the system level

- TAAF units are representative of production units

- TAAF process is integrated into the systems engineering development and test program to optimize the use of all assets, tests, and analyses.

- TAAF environments are based on worst case DRMP extremes, and normally include, as a minimum, vibration, temperature, shock, power cycling, input voltage variance, and output load

- TAAF is augmented by Failure Reporting and Corrective Action System to improve selected systems with a continuing history of poor reliability/performance

- HALT is performed at all hardware assembly levels as a development tool, and used as an alternative to TAAF to quickly identify design weaknesses and areas for improvement

- The mechanism of each failure, including stresses above the specification limits, is understood

- TAAF test resources should include between 2 to 10 Units-Under-Test (UUT), based on cost and complexity trade-off

- Ratio of TAAF test time at vibration and temperature extremes to total test times: $0.8 \leq$ Ratio ≤ 1.0

- Total calendar time (allocated and actual) to complete TAAF testing is approximately twice the number of test hours

- Test Time for each TAAF UUT is within 50% of the average time

Watch Out For

- Development programs with TAAF or HALT planned at the system level only

- TAAF planned or conducted in lieu of developmental/exploratory engineering tests

- TAAF testing conducted with a limited sample size and a limited number of test hours/cycles

- Use of Bayesian approaches to shorten TAAF test time and to estimate reliability when the a-priori data is questionable

- A tendency to focus on statistical measures associated with TAAF and HALT, rather than using test results to identify and correct design deficiencies

- TAAF UUT and test facilities that are not conditioned/groomed (burn in, screened, etc.) prior to test as planned for normal production

- Infant mortality failures are included in growth measurements

- The use of TAAF as a trial and error approach to correct a poor design

- The use of HALT to discover defects

4.14 Manufacturing Plan

The Manufacturing Plan describes all actions required to produce, test and deliver acceptable systems on schedule and at minimum cost. The materials, fabrication flow, time in process, tools, test equipment, plant facilities and personnel skills are described and integrated into a logical sequence and schedule of events.

Best Practice

- Identification, during design, of key product characteristics and associated manufacturing process parameters and controls to minimize process variations and failure modes

- FMECA of the manufacturing process during design for defect prevention

- Specified manufacturing process variability (e.g. Cpk) is within the design tolerances

- Variations of test and measuring equipment are accounted for when determining process capability

- Rapid prototyping for reduced cycle time from design to production (see Rapid Prototyping).

- Design For Manufacturing and Assembly to develop simplified designs

- Design for agile manufacturing to quickly adapt to changes in production rate, cost and schedule.

- Contingency planning for disruption of incoming parts, variations in manufacturing quantities, and changes in manufacturing capabilities

- Controlled drawing release system instituted (see Design Release)

- Process proofing/qualification (see Manufacturing Process Proofing/Qualification)

- Product/process changes that require qualification are defined

- Flowcharts of manufacturing processes at the end of EMD, validated at the start of LRIP

- Facilities, manpower, and machine loading for full rate production are validated during LRIP. Production readiness reviews performed on critical processes

- Subcontractor process capabilities integrated into the prime contractor's process capabilities

- Specific product tests and inspections replaced with Statistical Process Controls (SPC) on a demonstrated capable and stable process

- Closed loop discrepancy reporting and corrective action system, including customer and subcontractor discrepancies

- Post production support plan established and maintained for:

  - Repair capability

  - Obsolescence of tools, test equipment and technology

  - Loss of contractor expertise and vendor base, and

  - Time/cost to reestablish production line

Metrics Include:

- Measurable key characteristics and parameters are identified on drawings, work instructions and process specification

- SPCs (e.g., Cpk>1.33) are established for key characteristics

- Critical processes under control prior to production implementation

Watch Out For

- Total cost of "hidden factory" for non-conforming materials

- A deficient Materials Requirements Planning system

- Inadequate response planning for subcontractor design and production process changes

- Establishment of SPC for key processes without use of statistical techniques (e.g., Design of Experiments, Taguchi, QFD) or adequate run time to determine variability of the process when stable

- Operator self-checks without a process to verify integrity of the system

- Planning which permits production work-a-rounds and fails to emphasize scheduled production outputs.

4.15 Rapid Prototyping

Rapid Prototyping utilizes physical prototypes created from computer generated three-dimensional models to help verify design robustness as well as reduce engineering costs during production activities associated with faulty or difficult to manufacture designs. The use of these prototypes includes functional testing, producibility, dimensional inspection, assembly training, as well as tool pattern development.

Best Practice

- Rapid prototyping technology used in developing a product from concept to manufacturing

- Used to reduce design cycle time, iterate design changes, check fit and interfaces, calculate mass properties and identify design deficiencies

- Used in manufacturing producibility studies, proof of tooling and fixtures, training, and as a visualization aid in the design of the evolving product

- Virtual reality prototypes are analyzed using CAD tools and physical parts are fabricated from the CAD three dimensional drawings and data prior to production

Watch Out For

- Rapid prototyping without three dimensional CAD data for precise geometric representation

- Two-dimensional CAD surface model used in lieu of the more complete three dimensional solid model

- Rapid prototyping without a support structure to sustain the part in place while it is being generated

4.16 Manufacturing Process Proofing / Qualification

Manufacturing Process Proofing/Qualification ensures the adequacy of production planning, tool design, assembly methods, finishing processes and personnel training before the start of rate production. This is done in a time frame that allows for design and configuration changes to be introduced into the product baseline.

Best Practice

- Proofing simulates actual production environments and conditions

- "Proof of Manufacturing" models used to verify that processes and procedures are compatible with the design configuration

- First article tests and inspections included as part of process proofing

- Conforming hardware consistently produced within the cost and time constraints for the production phase

- Key processes are proofed to assure key characteristics are within design tolerances

- Process proofing must occur with:

  - A new supplier
  - The relocation of a production line
  - Restart of a line after a significant interruption of production
  - New or modified test stations, tools, fixtures, and products
  - Baseline and subsequent changes to the manufacturing processes
  - Special processes (non-testable/non inspectable)
  - Conversion of manual to automated line

Watch Out For

- Process proofing that does not include integration into higher assemblies to assure proper fit and function at the end item level

- Changes in subcontractor processes that occur without notifying the prime

- The use of Statistical Process Control (SPC) to qualify or validate the manufacturing process in lieu of first article tests and inspections

- The use of acceptance tests in lieu of process proofing or performance of first article tests and inspections

- Performance of first article tests and inspections only when contractually required

- Attempts to cite the warranty provisions rather than actually proofing the processes

- Overly ambitious schedule for qualification of new products/sources

4.17 Conformal Coating for Printed Wiring Boards

A conformal coating is a thin film applied to the surface of a Printed Wiring Board or other assembly which offers a degree of protection from hostile environments such as moisture, dust, corrosives, solvents and physical stress.

Best Practice

- Use trade studies to weigh the effects of conformal coating on long-term reliability, safety, and rework costs against potential savings in production and repair costs

- Conformal coating is used in environments where contaminants cannot be adequately controlled, including manufacturing or testing facilities

- Match the type of conformal coating to the configuration, maintenance concept and the use environment of what you want to coat

- Inspection techniques in place to verify uniformity and completeness of conformal coating coverage

- (See Table 4.17-1 for selected coating properties)

Watch Out For

- Conformal coating used to meet hermetic requirements, since conformal coating is not hermetic or waterproof

- Manufacturing and/or testing processes lacking a Failure Reporting and Corrective Action System and quality system to ensure that precautions against contaminants are effective, especially on assemblies without conformal coating

- The application of conformal coating to a non-coated assembly without first assessing the effects on circuit operating frequencies, mechanical stresses, thermal hot spots, etc. that may increase failure rates

- The use of assemblies without conformal coating that contain critical analog circuits and/or high-power circuits, possibly creating safety hazards

- The use of conformal coating that is not compatible with the repair philosophy

- The toxicity and environmental friendliness of conformal coating, including its by-products

- Inadequate surface preparation and condition prior to application of conformal coating

- Improper masking prior to conformal coating

Table 4.17-1. Conformal Coating Material Properties

Coating Properties	AR (Acrylic Resin)	UR (Urethane Resin)	ER (Epoxy Resin)	SR (Silicone Resin)	XY (Paraxylyene)
Nominal Thickness, Mils	1-3	1-3	1-3	2-8	0.6-1.0
Performance Under Humidity	Good	Good	Good	Good	Good
Resistance to Solvents	Poor	Good	Very Good	Fair	Excellent
Reparability	Excellent	Good	Poor	Fair	Poor
Application Characteristics	Excellent	Good	Fair	Fair	Fair
Volatile Organic Compound Exempt	Some	Some	Some	Some	All
Max. Continuous Operating Temp.	125C	125C	125C	200C	125C
Conveyor Processing Capability	Excellent	Poor to Fair	Poor to Fair	Poor to Fair	No

Source: Circuits Assembly, A Focus on Conformal Coating by Carl Tautscher, May 1997

4.18 Subcontractor Control

Reliance on subcontracting has made effective management of subcontractors critical to program success. Subcontractor Control includes the use of Integrated Product Teams, formal and informal design reviews, vendor conferences and subcontractor rating system databases.

Best Practice

- Subcontractor/supplier rating system with incentives for improved quality, reduced cost and timely delivery

- Flowdown of performance specification or detail Technical Data Package, depending on the acquisition strategy

- Subcontractors integrated into Integrated Product Teams to participate in the development of DRMP requirements

- Waiver of source and receiving inspections for subcontractors meeting certification requirements, depending on the product's criticality

- Subcontractor controls critical sub-tier suppliers

- Subcontractor notifies prime of design and process changes affecting key characteristics

- Metrics include subcontractor demonstrated process controls (e.g., Cpk > 1.33) for key characteristics

Watch Out For

- Procurement of critical material from an unapproved source

- Supplier performance rating does not consider the increased cost for defects discovered later in the prime's manufacturing process or after acceptance by the customer

- Subcontractor performance rating based primarily on cost, schedule and receiving inspection (vice performance requirements)

- Subcontractor process capability not verified

- Subcontractor decertification process is delinquent

4.19 Tool Planning

Tool Planning encompasses those activities associated with establishing a detailed, comprehensive plan for the design, development, implementation, and proof of program tooling. Tool planning is an integral part of the development process.

Best Practice

- Tools designed with CAD concurrent with product design

- Tool tolerances are at least 10% more restrictive than the hardware tolerances

- Measurement systems repeatability and reproducibility studies performed to establish the variability allowed to meet the key characteristic tolerances

- Tools are proofed, calibrated, certified and controlled

- Hard tooling validated prior to the start of production

- Tools are maintained with the aid of production statistical control charts

- Production tools are procured if the hardware is to be second sourced

- Minimize special tools and fixtures

- Metrics include:

  - Process capability Cpk > 1.33 for normal processes
  - Process capability Cpk > 1.67 for mission critical processes or for safety

Watch Out For

- Soft tooling used in production

- Calibration of tooling not traceable to a National standard and/or reference

- Master tooling not controlled

4.20 Special Test Equipment

Special Test Equipment (STE) is a key element of the manufacturing process used to test a final product for performance after it has completed in-process tests and inspections, final assembly and final visual inspection.

Best Practice

- STE is minimized

- ATE is developed for complex UUT, and considers test time limitations and accuracy

- STE accuracy/calibration must be traceable to known National measurement standard and/or references

- STE and applicable software are qualified, certified and controlled

- STE maintainability and maintenance concept defined concurrent with product design

- Life cycle functional and environmental profiles considered in STE design

- Design best practices are considered for critical STE

- Production demands are factored into STE design for reliability

- STE reliability target ≥ reliability of the system under test

- 4:1 minimum accuracy ratio between measurement levels (e.g., STE and UUT, standards and STE)

Watch Out For

- No fault repeatable loops

- STE software not validated

- STE production leads that impact increased rate production

- Root cause of STE discrepancies not understood

- STE false alarm rates

- STE not certified for acceptance testing

- Inadequate time between product CDR and STE delivery to support program schedule

4.21 Manufacturing Screening

Manufacturing Screening is a process for detecting in the factory, latent, intermittent, or incipient defects or flaws introduced by the manufacturing process. It normally involves the application of one or more accelerated environmental stresses designed to stimulate the product but within product design stress limits.

Best Practice

- Highly Accelerated Stress Screening (HASS) is performed as an environmental stress screen to precipitate and detect manufacturing defects

- HASS stress levels and profiles are determined from step stress HALT

- HASS precipitation screens are normally more severe than detection screens

- Product is operated and monitored during HASS

- The HASS screen effectiveness is proofed prior to production implementation

- HASS is performed with combined environment test equipment

- HASS stresses may be above design specification limits, but within the destruct limits, for example:

  - High rate thermal cycling

  - High level multi-axis vibration

  - Temperature dwells

- Input power cycling at high voltage

- Other margin stresses are considered when applicable to the product

• Alternative traditional environmental stress screening (ESS) guidelines for manufacturing defects may be in accordance with Tri-Service Technical Brief 002-93-08, "Environmental Stress Screening Guidelines," July 1993

• Parts Screening:

- 100% screening required when defects exceed 100 PPM

- 100% screening required when yields show lack of process control

- Sample screening used when yields indicate a mature manufacturing process

Watch Out For

• Inadequate fatigue life remaining in the product after HASS

• HASS stresses that only simulate the field environment

• Environmental conditions that exceed the material properties of the product

• HASS that does not excite the low vibration frequencies

4.22 Failure Reporting, Analysis and Corrective Action

Failure Reporting, Analysis and Corrective Action is a closed loop process in which all failures of both hardware and software are formally reported. Analyses are performed to determine the root cause of the failure, and corrective actions are implemented and verified to prevent recurrence.

Best Practice

• Failure Reporting, Analysis, and Corrective Action System (FRACAS) implementation is consistent among the Government, prime contractor and subcontractors

• FRACAS is implemented from the part level through the system level throughout the system's life cycle

• Criticality of failures is prioritized in accordance with their individual impact on operational performance

• All failures are analyzed to sufficient depth to identify the underlying failure causes and necessary corrective actions

• Subcontractor failures and corrective actions are reported to the prime

• Prime contractor is involved in subcontractor closeout of critical failures

• Failure database accessible by customer, prime contractor and subcontractors

- Failure Review Board is composed of technical experts from each functional area

- Test requirements established for Retest-OK/Can-Not-Duplicate (RTOK/CND) failures

Metrics Include:

- 100% of failures undergo engineering analysis

- 100% of critical failures undergo laboratory analysis

- Failure analysis and proposed corrective action are completed:

    - ≤ 15 (or what required) days for in-house analysis
    - ≤ 30 (or what required) days for outsourced analysis

- Feedback from the field to the factory should be in ≤ 30 days

Watch Out For

- Deferring FRACAS to the production phase

- No time limit for failure analysis and closeout

- Verification of corrective action not part of failure closeout

- Failures classified as random are not analyzed

- Failure analysis required only when repetitive failures occur

- Pattern of RTOK/CND failures

- Exclusion of test equipment, GFE and COTS/NDI failures from FRACAS

- Engineering and lab analysis not considering:

    - History of previous failures
    - Related circuit part failures
    - Temperature and other environmental conditions at failure
    - Workmanship precipitated failures correctable by design changes

- RF and other high energy part failures often results from test setup difficulties

- Backlog of failures to be analyzed in the laboratory

- Failure Review Board (FRB) and Quality Review Board (QRB) not integrated to review effectiveness of both functional and non-functional failures

- Failure closeouts dependent on FRB/QRB decisions

5.0 WARRANTY ANALYSIS

5.1 Warranty Management Strategies

It is crucial that organizations work on practices and processes to receive optimum benefits from warranty management. These include:

- Gaining insights on customer requirements for extended warranty claims, identifying the various sources of additional revenue streams
- Minimizing administrative costs by automating claims and settlement processes
- Including innovative process mechanisms like online real-time validation and claim submission to expedite processing time and enhance data quality
- Making on-time repairs possible through the easy availability of parts, enhancing customer satisfaction
- Enhancing the visibility of warranty information and introducing technical advancements in claim processing, reducing the possibility of fraud
- Monitoring to enhance quality issues through innovative methods
- Minimizing costs by launching supplier recovery programs

A proper implementation of these methods could help organizations explore and embrace the full potential of warranty management.

Price is the monetary amount a buyer pays a seller upon delivery of a product or service. In general, the price is the cost to produce the product or service plus a profit. This holds true for warranty pricing with the exception that there is no profit.

In establishing a warranty price, it must be both fair and reasonable. The price is defined as fair if both parties agree, given the quality and timeliness of performance. A cost is reasonable if, in its nature or amount, it does not exceed that which would be incurred by an ordinarily prudent person in the conduct of competitive business. The reasonableness of specific costs should be examined with particular care in connection with contracts that may not be subject to effective competition.

In addition to being fair and reasonable, costs included as part of a warranty price must be allocable and allowable. A cost is allocable if it is assignable or chargeable to a cost objective, in this case the warranty. A cost is allowable if it is reasonable, allocable, and generally accepted according to cost accounting standards principles and practices.

5.2 Statement of Contractor Warranty

Notwithstanding Government (User) inspection and acceptance of supplies and services furnished under this contract or any provision of this contract concerning the conclusiveness thereof, the Contractor guarantees that all ENDUNITS furnished under this contract shall be free from defects in design, material, and workmanship and shall operate in their intended environment in accordance with specifications, drawings and approved technical orders for the warranty period set forth herein. This warranty shall apply to all ENDUNITS furnished under this contract, including all option quantities procured, and all associated spares whether procured by the Government (User) or on loan from the Contractor.

Any ENDUNIT which fails shall be returned to the Contractor's designated repair facility at Government (User) expense. Such returns shall be either corrected or replaced at Contractor's expense, so as to operate in accordance with Statement of Work (SOW), CDRL, etc. Returned ENDUNIT as corrected or replaced and accepted by the Government (User) in accordance with approved repair verification test procedures shall be placed in secure storage, packaged and ready for issue. ENDUNIT shipped for correction on or before the expiration date of the warranty shall be covered under the terms of the warranty.

The Contractor shall not be obligated to correct or replace any ENDUNIT under the provisions of this warranty if loss or damage occurs by reason of (i) non-ENDUNIT induced fire, explosion, aircraft crash; (ii) submersion; (iii) acts of God such as flood, hurricane, tornado, or earthquake; and (iv) combat action.

The Contractor shall not be obligated under these warranty provisions for (i) repair or damage to warranted ENDUNIT caused by unauthorized maintenance by Government (User) personnel; (ii) repair of external physical damage caused by accidental or willful mistreatment by non-Contractor personnel; or (iii) repair of internal damage which, in the determination of the Administrative Contracting Officer (ACO), was caused by accompanying external damage.

The conditions specified in paragraphs except for natural disasters, apply only to loss or damage occurring on locations other than those owned or controlled by the Contractor or occurring while the ENDUNIT is not under Contractor's possession or custody. Under paragraph 2.3.2 only the repair of damage shall be the responsibility of the Government (User). All other expenses to receive, process and store a returned ENDUNIT shall be the responsibility of the Contractor as a part of his warranty obligations, unless there is clear and convincing evidence that the unit was returned solely because of the exclusion condition.

Specific cases wherein the Contractor and the ACO cannot agree on warranty coverage shall be referred to the Primary Contracting Officer (PCO) for final disposition. PCO dispositions that are not agreed to by the Contractor shall be subject to the "Disputes" language of the contract.

Specific cases wherein the ACO determines that the ENDUNIT is not covered within the terms of the warranty and is not correctable, the equipment shall be disposed of as directed by the USAF (User) Item Manager (IM) through the ACO. ENDUNIT disposed of, in accordance with this paragraph, or declared lost, may be replaced by the Government (User) with new ENDUNIT pursuant to the requirement entitled "Option for Increase Quantities." Replacement equipment shall continue to be warranted until the end of the Reliability Improvement Warranty (RIW).

The Contractor shall not be responsible under these terms and conditions for the correction of defects in Government (User) furnished property, except for defects in installation, unless the Contractor performs or is obligated to perform any modifications or other work on such property. In that event, the Contractor shall be responsible for correction of defects that result from the modifications or other work on such property.

In no event shall the Contractor be liable for special, consequential or incidental damages and nothing within the terms, conditions and requirements of this RIW provision shall be construed as being, or amounting to, special, consequential or incidental damages.

For purposes of this RIW, a failure is defined as any warranted ENDUNIT returned to the Contractor because of a failure indication, a malfunction and/or a reduction in the performance of the ENDUNIT below requirements of the contract specification.

5.3 Warranty Price Analysis Methods

Warranty price depends on two metrics: the type of warranty and the life distribution of the warranted item. The type of warranty determines the contractor repair obligation, whereas the life distribution determines the number of failures over the period of the warranty.

According to the *Analysis* of Warranty Cost Methodologies prepared by ARINC Research Corporation for the Naval Material Command, there are three methods for determining warranty price. These are:

- rule-of-thumb ratio,
- warranty cost estimating relationships (CERs)
- bottom-up accounting approach

5.3.1. *Rule-of-Thumb Ratio*

The rule-of-thumb is actually a warranty price ratio and is usually expressed as a percent of the unit production price per year. Applying the rule-of-thumb requires an extensive data base from which the ratio can be computed. Ideally, the ratio should be based on similar systems of comparable technology and complexity and with the same type of warranty provision. Research done by ARINC Corporation in 1985 showed that the rule-of-thumb ratio ranged from 0.2 to 12.6 percent. Performance and defects warranties were on the low end, while reliability improvement warranties tended to be on the high end. The rule-of-thumb ratio is best applied in the early phases of system development and acquisition or as a cross check in the later phases.

5.3.2 *Warranty Cost Estimating Relationships*

Just as with the rule-of-thumb ratio, warranty CERs require an extensive data base and an experienced analyst to manipulate the data base to produce a reliable, dependable result. In the broadest sense, a CER is a mathematical expression relating cost as a dependent variable to one or more independent cost-driving variables. A warranty CER can be classified into one of two types: factor based or regression based.

A factor based CER is a logical expression relating cost to cost. The factors on the right hand side of the equation are added, subtracted, multiplied, and divided to yield the resultant cost. The following equation is a factor based CER developed by TASC:

$$W = P + C_W + \left(\frac{Q_T U t_W}{MTBF_a}\right) C_r + I[MTBF_a] + D_t$$

where

W = Warranty price
P = Profit = 0
C_W = Contractor fixed costs
Q_T = Total number of systems to be delivered
U = Usage rate in operating time per calendar period
t_W = Duration of the warranty period
$MTBF_a$ = Achieved MTBF at the end of the warranty period
C_r = Cost to the contractor per unit repair
$I[MTBF_a]$ = Cost of improvements to achieve $MTBF_a$
D_t = Cost of damages for not meeting the turnaround time requirement

This CER is intended to estimate the price of a reliability improvement warranty. The last two terms, $I[MTBF_a]$ and D_t, could be dropped to generalize the expression for non-RIW warranties.

Regression based CERs are based on the premise that a statistical relationship exits between warranty price and price-determining variables. As an example, it is reasonable to assume that warranty price would vary with the number of system failures. Using historical data, regression analysis can be used to develop the CER. Regression based CERs rely on an extensive, well-developed data base.

A simple linear regression based CER may be of the form:

$$W = aX + b$$

where

W = Warranty price
X = Number of failures

a = Regression derived parameter associated with the variable cost component of warranty price
b = Regression derived parameter representing the fixed cost component of warranty price

A simple non-linear regression based CER may be of the form:

$$W = aX^b$$

where

W = Warranty price
X = Number of failures
a = Regression derived parameter
b = Regression derived parameter

This expression can be linearized using a logarithmic transformation as follows:

$$\ln W = \ln a + b \ln X$$

where

ln W = Natural log of W
ln a = Natural log of a
ln X = Natural log of X
W = Warranty price
X = Number of failures
a = Regression derived parameter
b = Regression derived parameter

In the two examples thus far presented, simple regression using one independent and one dependent variable has been used. For many pricing applications this may be sufficient if price can be accurately determined using a single, key cost driver. In other applications, a single independent variable may not be adequate to predict price reliably. In these instances, it is useful to generalize the simple regression techniques discussed above and add additional independent variables. This more general form of regression is called multiple regression since 'multiple' independent variables are employed. Therefore, the linear mathematical form of the equation showing the relationship between the single dependent variable and the multiple independent variables is:

$$W = a + b_1X_1 + b_2X_2 + \ldots + b_nX_n$$

in the linear case, where

W = Warranty price
a = Regression derived parameter
$b_1, \ldots, b_n$ = Regression derived parameters
$X_1, \ldots, X_n$ = Independent variables used to predict warranty price
n = Number of independent variables

5.3.3 Bottom-Up Accounting Approach

The third method for estimating warranty price employs the bottom-up accounting approach. The basic procedure to be followed in applying the model is to identify all warranty cost elements and, in an accounting manner, estimate the cost of the factors involved and add them to arrive at the total cost of the warranty. That is, the total is equal to the sum of the parts. Estimation of the partial costs may involve the use of rules-of-thumb or cost estimating relationships. The bottom-up approach depends on a detailed examination of the events that could occur during the warranty period, how often they may occur, and what their cost may be.

One way to begin an application of the bottom-up approach would be to divide warranty price into two components: fixed costs and variable costs. Therefore,

$$W = FC + VC$$

where

W = Warranty price
FC = Fixed costs
VC = Variable costs

Fixed costs consist of facilities to store and repair warranted items, equipment to perform warranty repairs, certain management and administration costs, and overhead. Variable costs consist of direct labor to perform the repairs, parts and material for the repairs, handling costs, and the remainder of the management and administrative costs. If included in the warranty price, second destination transportation could also be a variable cost. Second destination transportation includes the cost of packaging and shipping components between the operational locations and the repair site. Second destination transportation is usually a government responsibility, however this is negotiable between the contractor and the government. At any rate, the analyst must ensure that second destination transportation cost is included in the analysis.

When determining the variable cost per repair, several items of information are necessary. The first is the contractor touch labor rate. This varies by skill level and geographical location of the repair facility. Next is the repair time most often expressed as mean time to repair or MTTR. The repair time will vary considerably depending on the failure mode. Then, the labor rate per hour times the repair time, also in hours, gives the variable labor cost for a single repair.

The number of failures is determined by the mean time between failure (MTBF). As with repair time, the MTBF is determined by the failure mode. A common assumption is that the number of failures can be modeled by a Poisson process. That is the number of failures follow a Poisson probability distribution and the time between failures is distributed exponentially. If this is the case, then the expected number of failures in time t can be expressed as:

$$X_t = \sum_{k=0}^{\infty} \frac{(\lambda t)^k}{k!} e^{-\lambda t}$$

where

X_t = Expected number of failures in time t
λ = Average arrival rate of failures (i.e. failures per hour)
MTBF = mean time between failures = $1/\lambda$

Variable repair cost can then be estimated by the equation

$$VC = ((LC)(MTTR) + (MC) + (HC) + (AC)) \frac{T}{(MTBF)}$$

where

VC = Variable cost
LC = Labor cost per unit of time
MTTR = Mean time to repair
MC = Material and repair part cost
HC = Handling cost per failure
AC = Administrative and management cost per failure
T = Warranty period or time
MTBF = Mean time between failure

The bottom-up accounting approach, also known as 'engineering build-up' or 'grass roots estimating', may more than likely begin with the work breakdown structure (WBS). The bottom-up accounting approach would normally be used just prior to or during the production phase of the program cycle. At this stage of the program, the end item configuration is stabilized and test results are available.

5.3.4 Sparing Analysis

Shown below is the method to calculate the number of spare Line Replacement Units (LRU) that should be initially available to ensure a preselected probability (confidence) that a spare is available

$$CL \leq \sum_{k=0}^{S} \frac{(n\lambda R)^k}{k!} e^{-n\lambda R}$$

where: n = Number of LRU in service

λ = Failure rate per hour

R = Repair time in hours

CL = Confidence Level

S = Minimum number of spares required

An example is shown how this equation can be used to determine the number of spares that must be available taking into account the repair time, the failure rate of the LRU, number of units in service, and a given level of confidence

Case 1: Failure Rate = 100 failures per million hours; Mean Time to Repair = 4 hours; LRU in service = 1000; If spares required = 1, Confidence Level = 93.84 %

Case 1: Failure Rate = 100 failures per million hours; Mean Time to Repair = 4 hours; LRU in service = 1000; If spares required = 2, Confidence Level = 99.21 %

Case 1: Failure Rate = 100 failures per million hours; Mean Time to Repair = 4 hours; LRU in service = 1000; If spares required = 3, Confidence Level = 99.92 %

Case 1: Failure Rate = 100 failures per million hours; Mean Time to Repair = 4 hours; LRU in service = 1000; If spares required = 4, Confidence Level = 99.99 %

The calculator in the Websites given below calculates the sparing needs to satisfy a set of given conditions

[http://src.alionscience.com/cgi-src/formdraw.pl?Change2=6]
[http://reliabilityanalyticstoolkit.appspot.com/poisson_spares_analysis]

6.0 GLOSSARY OF TERMS

Accessibility
A measure of the relative ease or difficulty of entry to various areas of an item or system for operation or maintenance. Includes consideration of free space or lack thereof surrounding an item to allow or restrict the technician/operator's approach to necessary servicing, maintenance, and operating locations, e.g. switches, connectors, fasteners, output devices and displays.

Affordability.
A subjective decision that a system's anticipated total life cycle costs are sufficiently cost effective over the system's expected service life to justify its research, development, procurement, operation, and maintenance in a given operating environment. (See Cost as an independent variable and Design-to-cost).

Availability.
A measure of the degree to which an item is in an operable state at any given (random) time. increasing failure rate during the wear out period. Plotting failure rate vs. time yields a U-shaped curve reminiscent of a bathtub.

Capability.
A measure of the system's ability to achieve mission objectives, given the system condition during the mission.

Commonality.
A quality which applies to material or systems possessing like and interchangeable characteristics. Having interchangeable repair parts and/or components. Applies to consumable items interchangeable without adjustment.

Compatibility.
The ability of two or more operational items/systems to exist or function as elements of a larger operational system or operational environment without mutual interference.

Configuration.
A collection of an item's descriptive and governing characteristics, which can be expressed in functional terms (what performance the item is expected to achieve) or physical terms (what the item should look like and consist of when built).

Configuration Item (CI).
An aggregation of hardware, firmware, or computer software or any of their discrete portions which satisfies an end use function and is designated by the Government for separate configuration management. Configuration items may vary widely in complexity, size, or type. Any item required for logistic support and designated for separate procurement is a configuration item.

Configuration Management (CM).
Technical and administrative direction and surveillance actions taken to identify and document functional and physical characteristics of an item; to control changes to an item and its characteristics; and to record and report the change processing and implementation status of an item.

Consumable.

Any item not specifically identified as controlled equipage or spare parts that is used up or loses its identity in normal usage. Examples are administrative or housekeeping items, general purpose hardware, some bit or piece parts, and common tools.

Contract.

A mutually binding legal relationship obligating the seller to furnish the supplies or services (including construction) and the buyer to pay for them. It includes all types of commitments that obligate the Government to an expenditure of appropriated funds and that, except as otherwise authorized, are in writing. In addition to bilateral instruments, contracts include (but are not limited to) awards and notices of awards; job orders or task letters issued under basic ordering agreements; letter contracts; orders, such as purchase orders, under which the contract becomes effective by written acceptance or performance; and bilateral contract modifications.

Criticality.

A measure of the seriousness of the consequences of a failure mode along with its probability of occurrence.

Damage Tolerance.

The ability of an item to continue to perform its designed function without repair for a specified period in spite of the presence of flaws or damage caused by initial manufacture, the operating environment (vibration, thermal stress, etc.), abuse, or enemy action.

Depot Level Maintenance.

Maintenance performed on material requiring major overhaul or a complete rebuild of parts, subassemblies, and end items, including the manufacture of parts, modification, testing, and reclamation as required. Depot maintenance serves to support lower categories of maintenance by providing technical assistance and performing that maintenance beyond their responsibility. Depot maintenance provides stocks of serviceable equipment by using more extensive facilities for repair than are available in lower level maintenance activities.

Design Interface.

The process of including reliability, maintainability, supportability, and other logistics concerns in the design of a product or technology, beginning in the concept exploration and definition phase or earlier, and continuing throughout the product's acquisition life cycle. One of the principal elements of ILS.

Design-to-Cost (DTC).

Management concept wherein rigorous cost goals are established during development and the control of systems costs (acquisition, operating, and support) to these goals is achieved by practical tradeoffs between operational capability, performance, costs, and schedule. Cost, as a key design parameter, is addressed on a continuing basis and as an inherent part of the development and production process. .
The act of getting rid of excess, surplus, scrap, or salvage property under proper authority. Disposal may be accomplished by, but not limited to, transfer, donation, sale, declaration, abandonment, or destruction. Demilitarization may be required prior to or as part of the disposal process.

Durability.

The ability to resist wear, cracking, corrosion, deterioration, thermal degradation, etc. while continuing to function as designed under specified conditions for a specified period.

Effectiveness.
The extent to which the goals of the system are attained, or the degree to which a system can be elected to achieve a set of specific mission requirements.

Electromagnetic Compatibility (EMC).
(1) A measure of equipment tolerance to an electromagnetic field.

(2) The ability of a device to function satisfactorily in its electromagnetic environment without introducing intolerable disturbance to the environment (or to other devices).

Electromagnetic Interference (EMI).
Engineering term used to designate interference in a piece of electronic equipment caused by another piece of electronic or other equipment.

Environmental Stress Screening (ESS).
A series of tests designed to remove latent part and manufacturing defects through application of environmental stimuli prior to fielding of the equipment. Random vibration and thermal cycling are the most common environmental stress tests used.

Failure.
The event in which any part of an item does not perform as required by its performance specification.

Failure Analysis.
The systematic examination of a failed item, its application, design, construction, and documentation to identify the failure mode and determine the basic failure mechanism and the course of failure propagation.

Failure Mode and Effects Analysis (FMEA).
A reliability evaluation and design review technique that examines potential failure modes within a system in order to determine the effects of failures on system performance.

Failure Rate.
Total number of failures divided by the total number of life units (operating hours, cycles, etc.) experienced during a particular measurement interval under specified conditions.

Fault Isolation.
The process of determining the location of a fault to the extent necessary to effect a repair.

Form, Fit, and Function Data.
Technical data pertaining to items, components, or processes for the purpose of identifying source, size, configuration, mating, and attachment characteristics, functional characteristics, and performance requirements.

Inherent Availability.
Availability of a system with respect only to operating time and corrective maintenance. It ignores standby and delay times associated with preventive maintenance as well as administrative and logistics down time.

Inherent R&M Value.
Any measure of reliability or maintainability that includes only the effects of item design and installation, and assumes an ideal operating and support environment.

Initial Spares.
Items procured for logistics support of a system during its initial period of operation.

Inspection.
The examination and testing of supplies and services (including, when appropriate, raw materials, components, and intermediate assemblies) to determine whether they conform to specified requirements.

Integrated Diagnostics.
An initiative for delivering weapon systems designed for ease of maintenance (with built-in diagnostics) with less test equipment and fewer maintenance specialists. Suggested by industry, it enhances military capabilities by increasing survivability of the support structure and by reducing the logistics tasks which could degrade unit mobility.

Integrated Logistics Support (ILS).
A disciplined, unified, and iterative approach to the management and technical activities necessary to integrate support considerations into system and equipment design; develop support requirements that are related consistently to readiness objectives; acquire the required support; and provide the required support during the operational phase at minimum cost.

Integrated Logistics Support Elements.
The principal elements of ILS include: Maintenance Planning; Manpower and Personnel; Supply Support; Support Equipment; Technical Data; Training and Training Support; Computer Resources Support; Facilities; Packaging, Handling, Storage, and Transportation; and Design Interface.

Intermediate Level Maintenance.
That level which maintains/repairs items for which the organizational level is incapable, but which do not have to be sent to the depot level for major work. Intermediate maintenance is the responsibility of designated maintenance activities for direct support of the using organization. Its phases normally consist of calibration, repair or replacement of damaged or unserviceable parts, components or assemblies; the emergency manufacture of unavailable parts, and providing technical assistance to using organizations.

Life Cycle Cost.
The total cost to the government of acquisition and ownership of a system or item over its useful life. It includes the cost of development (RDT&E), acquisition (production), deployment, operations and support (including manpower), modification (if applicable), and disposal (if applicable).

Line Replaceable Unit (LRU).
An essential support item removed and replaced at organizational level to restore an end item to an operationally ready condition. (Also called Weapon Replacement Assembly and Module Replaceable Unit).

Logistic Interoperability.
A form of interoperability in which the service to be exchanged is assemblies, components, spares, or repair parts. Logistic interoperability will often be achieved by making such assemblies components, spares, or repair parts interchangeable, but can sometimes be a capability less than interchangeability when a degradation of performance or some limitations are operationally acceptable.

Logistics.
The science of planning and carrying out the movement and maintenance of forces. In its most comprehensive sense, those aspects of military operations which deal with:

  (a) design and development, acquisition, storage, movement, distribution, maintenance, evacuation, and disposition of materials;

  (b) movement, evacuation, and hospitalization of personnel;

  (c) acquisition or construction, maintenance, operation, and disposition of facilities;

  (d) acquiring or furnishing of services.

Logistics Support Analysis.
The selective application of scientific and engineering efforts undertaken during the acquisition process, as part of the systems engineering process, to assist in: causing support considerations to influence design; defining support requirements that are related optimally to design and to each other; acquiring the required support; and providing required support during the operational phase at minimum cost. Often strongly associated with MIL-HDBK-502A, Product Support Analysis.

Logistics Support Analysis Record (LSAR).
A formal tool under MIL-HDBK-502A that uses records/forms to document operations and maintenance requirements, RAM, task analyses, technical data, support/test equipment, facilities, skill evaluation, supply support, ATE and TPS, and transportability. LSAR is the basis for training, personnel, supply provisioning and allowances construction, support equipment acquisition, facilities construction and preparation, and for maintenance -- preventative and corrective.

Maintainability.
A measure of the ability of an item to be retained in or restored to a specified condition when maintenance is performed using prescribed procedures and technician skill levels.

Maintenance.
The combination of all technical and administrative actions intended to retain an item in, or restore it to, a state in which it can perform its required function. Includes tests measurements, replacements, adjustments, and repairs. Software maintenance includes program copying and program improvement. Maintenance may be either corrective or preventive. Preventive maintenance performed at designated points in an item's life.

Mission Critical System.
A system whose operational effectiveness and operational suitability are essential to successful completion or to aggregate residual combat capability. If this system fails, the mission likely will not be completed. Such a system can be an auxiliary or supporting system, as well as a primary mission system.

Operational Effectiveness.
The overall degree of mission accomplishment of a system when used by representative personnel in the environment planned or expected (e.g., natural, electronic, threat etc.) for operational employment of the system considering organization, doctrine, tactics, survivability, vulnerability, and threat (including countermeasures, initial nuclear weapons effects, nuclear, biological, and chemical contamination (NBCC) threats).

Organizational Level Maintenance.
Maintenance which is the responsibility of and performed by the using organization on its assigned equipment. Its phases normally include inspecting, servicing, lubricating, adjusting and replacing plug-in modules and other parts, minor assemblies, and subassemblies with relatively short isolation and replacement times.

Program Management.
The process whereby a single leader and team are responsible for planning, organizing, coordinating, directing and controlling the combined efforts of participating/assigned civilian and military personnel and organizations in accomplishment of program objectives. A special management approach used to provide centralized authority and responsibility for the management of a specific defense acquisition program or programs. Program management provides a single point of contact as the major force for directing the system through development, production and deployment.

Project.
(1)   Synonymous with program in general usage.

(2)   Specifically, a planned undertaking having a finite beginning and ending, involving definition, development, production, and logistics support of a major weapon or weapon support system or systems. A project may be the whole or a part of a program.

Provisioning.
The process of determining and acquiring the range and quantity (depth) of spares and repair parts, and support and test equipment required to operate and maintain an end item of material for an initial period of service. Usually refers to first outfitting of a ship, unit or system.

Redundancy.
The existence of two or more means (not necessarily identical) for accomplishing a given system function. In general, redundancy improves system reliability.

Reliability.
The probability that an item will perform its intended function for a specified time under stated conditions. The ability of a system and its parts to perform its mission without failure, degradation, or demand on the support system.

Reliability Growth.
Improvement in reliability caused by the successful correction of deficiencies in an item's design or manufacture by way of an iterative process.

Retrofit.
Adding a new type of equipment to the configuration of operational systems or installing equipment in production systems which were delivered without such equipment.

Risk Management.
All actions taken to identify, assess, and eliminate or reduce risk to an acceptable level in selected areas (e.g., cost, schedule, technical, and producibility); and the total program.

Service Life.
Quantifies the average or mean useful life of the item. There is no general formula for the computation. Service life refers to the mean time between overhauls, or to the mandatory replacement time, or to the total time that an item retains usefulness in respect to the weapon system it supports.

Service life may be predicted based on reliability models for new or modified systems or it may be derived from historical data for existing systems.

Shelf Life.
The length of time an item can be kept in storage under specified conditions and still meet specified requirements. Also called storage life.

Shop Replaceable Unit (SRU).
An essential support item removed and replaced at the intermediate level to restore an end item or line replaceable unit to an operationally ready condition.

Single Point Failure.
The failure of a system caused by the failure of a single item or component which is not compensated for by redundancy provisions in design or by alternative operating procedure(s).

Specification.
A document intended primarily for use in procurement, which clearly and accurately describes the essential technical requirements for items, materials or services including the procedures by which it will be determined that the requirements have been met. Specifications may be prepared to cover a group of products, services, or materials, or a single product, service or material, and are general or detail specifications.

Statement of Work (SOW).
That portion of a contract which establishes and defines all non-specification requirements for contractors efforts either directly or with the use of specific cited documents.

Statistical Process Control (SPC).
The use of statistical techniques such as control charts to analyze a process or its outputs so as to take appropriate actions to achieve and maintain a state of statistical control and to improve the process capability.

Supportability.
The design characteristics of a system or subsystem which permit or improve/simplify servicing, preventive and corrective maintenance; improve interoperability with other services or allies; reduce storage requirements; increase shelf life; minimize support equipment needs; reduce transportation needs; reduce support manpower requirements; and/or reduce the support costs while allowing the system to meet its peacetime and wartime mission requirements. (See System).
The degree to which a device, equipment, or weapon system is open to effective attack due to one or more inherent weakness. Susceptibility is a function of operational tactics, countermeasures, probability of enemy fielding a threat, etc. Susceptibility is considered a subset of survivability.

System.
(1) The organization of hardware, software, material, facilities, personnel, data, and services needed to perform a designated function with specified results, such as the gathering of specified data, its processing, and delivery to users.

(2) A combination of two or more interrelated equipments (sets) arranged in a functional package to perform an operational function or to satisfy a requirement.

System Dependability.
The probability that the hardware and software will perform successfully during one or more required sequences of a mission, given the hardware and software status at the start of the mission (availability).

Systems Engineering.
The application of scientific and engineering efforts to

(a) transform an operational need into a description of system performance parameters and a system configuration through the use of an iterative process of definition, synthesis, analysis, design, test, and evaluation;

(b) integrate related technical parameters and ensure compatibility of all physical, functional, and program interfaces in a manner that optimizes the total system definition and design;

(c) integrate reliability, maintainability, safety, survivability, human, and other such factors into the total engineering effort to meet cost, schedule, and technical performance objectives.

Test.
Any program or procedure which is designed to obtain, verify, or provide data for the evaluation of: research and development (other than laboratory experiments); progress in accomplishing development objectives; or performance and operational capability of systems, subsystems, components, and equipment items.

Test and Evaluation (T&E).
Process by which a system or components are compared against requirements and specifications through testing. The results are evaluated to assess progress of design, performance, supportability, etc. There are three types of T&E-Development (DT&E), Operational (OT&E), and Production Acceptance (PAT&E) -- occurring during the acquisition cycle. DT&E is conducted to assist the engineering design and development process and to verify attainment of technical performance specifications and objectives. OT&E is conducted to estimate a system's operational effectiveness and suitability, identify needed modifications, and provide information on tactics, doctrine, organization, and personnel requirements. PAT&E is conducted on production items to demonstrate that those items meet the requirements and specifications of the procuring contracts or agreements. OT&E is further subdivided into two phases-Initial operational (IOT&E) and Follow-on Operational (FOT&E). IOT&E must be conducted before the production decision (Milestone III) to provide a credible estimate of operational effectiveness and suitability. Therefore, IOT&E must be conducted on a system as close to a production configuration as possible, in an operationally realistic environment, by typical user personnel. FOT&E is conducted on the deployed system to determine if operational effectiveness and suitability is, in fact, being attained.

Testability.
A design characteristic which allows the status of a unit to be confidently determined in a timely manner.

Three-Level Maintenance Concept.
The maintenance system consisting of organizational, intermediate, and depot maintenance levels, each of which has more repair capability and authority than the previous level. No longer the preferred maintenance philosophy.

Transportability.

The capability of materiel to be moved by towing, self-propulsion, or carrier through any means, such as railways, highways, waterways, pipelines, oceans, and airways. (Full consideration of available and projected transportation assets, mobility plans and schedules, and the impact of system equipment and support items on the strategic mobility of operating military forces is required to achieve this capability.)

Uptime Ratio.

Measure of operational availability and dependability that includes the combined effects of item design, installation, quality, environment, operation, maintenance, repair, and logistic support. Up time divided by active time.

Utilization Rate.

The planned or actual number of life units (operating hours, cycles, etc.) expended or missions attempted during a specified time interval.

7.0 INFORMATION SOURCES USEFUL TO "AVAILABILITY"

DOD 3235.1-H. 1982. Test and Evaluation of System Reliability, Availability, and Maintainability, A primer, March 1982. Washington, DC: Office of the Under Secretary of Defense for Research and Engineering

NEOS Optimization Guide provides information about the field of optimization and many of its sub-disciplines. The focus of the content is on the resources available for solving optimization problems, including the solvers available on the NEOS Server. [*https://neos-guide.org/Optimization-Guide*]

- Introduction to Optimization: provides an overview of the optimization modeling and solution process
  [*https://neos-guide.org/content/optimization-introduction*]

- Types of Optimization Problems: provides some guidance on classifying optimization problems
  o alphabetical list of optimization problem types
  o graphical taxonomy of optimization
- Optimization Algorithms: provides information on various solution methods along with links to software tools that implement the method
  [*https://neos-guide.org/algorithms*]

- FAQs: answers to frequently asked questions on Linear Programming and Nonlinear Programming
- Optimization Resources: links to blogs, communities, conferences, and publications related to optimization

The Handbook of Reliability Prediction Procedures for Mechanical Equipment (NSWC-06) was originally released in the 1980's by the Naval Surface Warfare Center. The development of this handbook was coordinated with the military, industry and academia. It was developed to present a proposed methodology for predicting the reliability of mechanical equipment
[*http://www.reliabilityeducation.com/intro_nswc.html*]

Intricacies of System of Systems Operational Availability and Logistics Modeling 18th Annual Systems Engineering Conference October 26 – 29, 2015
[*https://ndiastorage.blob.core.usgovcloudapi.net/ndia/2015/system/17994_Carter.pdf*]

A "Design for Availability" Methodology for Systems Design and Support, Taoufik Jazouli, Doctor of Philosophy, 2011 Directed By: Professor, Peter Sandborn, Mechanical Engineering
[*https://drum.lib.umd.edu/bitstream/handle/1903/12217/Jazouli_umd_0117E_12653.pdf;sequence=1*]

Department of Defence Reliability, Availability, Maintainability and Cost Rationale Report Manual, June 1, 2009.

This manual describes the development of the RAM and Cost Rationale Report (hereafter referred to as RAM-C Report). The guide was written to help capability document requirements writers and their supporting engineering organizations think through the top-level sustainment requirements for RAM-C early in the requirements generation and refinement phases of a program to ensure the system is sustainable and affordable throughout its life cycle. *[https://www.acq.osd.mil/se/docs/DoD-RAM-C-Manual.pdf]*

Ministry of Defence Defence Standard 00-45 Part 3 Issue 1 Publication Date 14 April 2006.
Using Reliability Centred Maintenance to Manage Engineering Failures Part 3 Guidance on the Application of Reliability Centred Maintenance. RCM is essentially a formal application of engineering logic supported by experience and sound engineering judgement. If applied correctly, it will result in a recommendation for the most cost-effective engineering failure management programme which will ensure assets continually meet required performance standards. It also helps to demonstrate that it is exercising a "duty of care" towards its employees and the environment as required by national and international legislation

OPNAV Instruction 3000.12A Operational Availability Of Equipments And Weapons Systems

This provide policy regarding Operational Availability (A_o) as a primary measure for readiness of naval systems, subsystems, and equipment. Provide definitions and equations for calculating A_o and identifying sources of data for calculating and monitoring A_o. *[https://doni.documentservices.dla.mil/Directives/03000%20Naval%20Operations%20and%20Readiness/03-00%20General%20Operations%20and%20Readiness%20Support/3000.12A.pdf]*

FAA-HDBK-006A January 7, 2008 Federal Aviation Administration Handbook

This contains Reliability, Maintainability, and Availability (RMA) Handbook. This handbook covers the development of reliability, maintainability and availability (RMA) requirements for the National Airspace System (NAS). This document will guide Units and acquisition managers in preparing procurement packages for major system acquisitions. RMA-related sections of these packages include Information for Proposal Preparation, System-Level Specifications, Statements of Work, and Data Item Descriptions. The handbook not only establishes RMA contractual requirements but also recommends comprehensive steps to ensure that fielded systems successfully comply with them. It provides guidance to help managers reduce NAS-Level requirements to levels of detail and characteristics that can readily be monitored and verified. Additionally, it recommends procedures to help managers evaluate proposals, monitor design development, and conduct effective tests and verifications. *[http://www.tc.faa.gov/its/worldpac/standards/faa-hdbk-006b.pdf]*

Department of Defense Reliability, Availability, Maintainability, and Cost Rationale Report Manual

This manual describes the development of the RAM and Cost Rationale Report (hereafter referred to as RAM-C Report). The guide was written to help capability document requirements writers and their supporting engineering organizations think through the top-level sustainment requirements for RAM-C early in the requirements generation and refinement phases of a program to ensure the system is sustainable and affordable throughout its life cycle.

The purpose of this manual is threefold:

1. Provide guidance in how to develop and document realistic sustainment Key Performance Parameter (KPP)/Key System Attribute (KSA) requirements and

related supporting rationale

2. Provide guidance so the acquisition community understands how the requirements must be measured and tested throughout the system life cycle

3. Describe desired processes for the Office of the Under Secretary of Defense, Joint Staff, and other stakeholders to interface with Services and programs when developing the susminment requirements.

Use of the processes outlined in this document will assist in assessing RAM-C for the alternatives considered in the Analysis of Alternatives and articulating the requirements and the supporting rationale in the Capability Development Document and Capability Production Documents and the Life Cycle Sustainment Plan.
[https://www.dsiac.org/sites/default/files/policy-standards/DoD-RAM-C-Manual%202009-06-01.pdf]

Prognostic Health Management (PHM)

PHM is a methodology that permits the assessment of the reliability of a system under its actual application conditions, and exercises necessary management actions. Prognostics is the process of predicting the future reliability of a product by assessing the extent of deviation or degradation of a product from its expected normal operating conditions. Health monitoring is aprocess of measuring and recording the extent of deviation and degradation from a normal operating condition.
[https://www.prognostics.umd.edu/]

Methodology for Estimation of Operational Availability as Applied to Military Systems:

Operational availability (Ao) is an important consideration during the evaluation of system effectiveness and sustainability. Ao is sometimes specified as an attribute within military requirements documents at the discretion of the proponent. Recently, however the Chairman the Joint Chiefs of Staff Manual (CJCSM) 3170C mandated the establishment of materiel availability as a sustainment Key Performance Parameter (KPP). KPPs are defined to be those attributes of a system that are considered critical or essential to the development of an effective military capability. However, test and evaluation of availability is problematic because it is highly dependant on the response and delay times associated with the maintenance and logistics support structures, which are not normally in place prior to fielding. This often leads to the evaluation of Ao via analysis or simulation—measuring the systems reliability and maintainability characteristics—and applying an estimate of the effect of the logistic support system. This article provides a brief background of Ao as well as a comparison of several methodologies for measuring and estimating Ao. Although KPPs are required by CJCSM 3170C to be testable, it is clear that it is necessary in most cases to measure the inherent reliability and maintainability of an item and to apply modeling and/or simulation techniques to evaluate the actual Ao. The equations and methodologies in the article describe the most common of those techniques, as well as their limitations and shortcomings.
[http://www.dtic.mil/docs/citations/ADA518378]

An Algorithm to Partition the Operational Availability Parts of an Optimal Provisioning Strategy

The implementation of an appropriate and cost-effective sparing provisioning strategy depends on the choice of an adequate mathematical model. The model chosen should provide optimal cost-effective provisioning as well as other applicable factors and should emphasize the following aspects for a total system evaluation: (i) identification of system parameters, and (ii) interactions of system parameters for optimal evaluation. The different parts that make up the system operational availability (A_O) must be identified. If an algorithm could be developed to represent the different parts of A_O, the individual contributions of these could be identified and used to determine an optimal provisioning strategy or other factors that influence the overall operational availability. From the basic RMA (Reliability, Maintainability, Availability) concepts, analytical derivations, and empirical example, an algorithm can be used to express the operational availability as a function of all its parts. The mathematical model developed to partition ail of the individual availabilities of the system A_O is a simple and direct way to use the calculated values of the individual availabilities. By partitioning system operational availability into all its parts, it can be easily evaluated, trade studies can be made, and it can be optimized for a cost-effective provisioning strategy. This paper presents an algorithm that makes it possible to partition easily, directly, and concisely by using the already calculated individual availabilities

[http://ieeexplore.ieee.org/document/744137/?reload=true]

Predicting Operational Availability for Systems with Redundant, Repairable Components and Multiple Sparing Levels.

The Strategic Petroleum Reserve (SPR) is a multi-mission project required by law to maintain a prescribed degree of readiness and a mandated performance criteria. The prediction of operational availability is essential to determine operational readiness to satisfy mission requirements. This is accomplished through, the use of availability models utilizing a reliability block diagram (RED) of mission critical components. The RED model calculations incorporate sparing criteria and components using a multi-state model of the operation. Individual component data include: capacity, meantime-between-failure (MTBF), and mean-down-time (MDT) assuming repairable components and instantaneous switching. An accurate site model addresses all of these concerns and provides a good prediction of operational availability. An example of a system without a spare and the same system with a spare is presented to illustrate one method of incorporating sparing into the prediction of operational availability.

[http://ieeexplore.ieee.org/document/500679/]

A Nonexponential Approach to Availability Modeling.

Most current state-of-the-art availability models are based on continuous-time Markov chains. This involves restrictive assumption about the probability distribution for both failure times and repair times being exponential. In many situations, the exponential distribution is not applicable for failure times and/or repair times. A general approach for calculating instantaneous availability is presented. It is applicable to systems or subsystems which are assumed to be returned to approximately their original state upon the completion of repair. . The first case study is a validation study since the uptimes and downtimes are both assumed to follow an exponential distribution. In this case, an analytical result for A(t) can be obtained. Thus, the results for the analytical approach and the proposed approach can be compared. An analysis of the results shows the proposed approach to be very reasonable. In the second case study, the uptimes are assumed to follow a Weibull distribution while the downtimes have a lognormal distribution

[https://www.researchgate.net/publication/3647659_A_nonexponential_approach_to_availability_modeling]

Operational Availability Modeling for Risk and Impact Analysis

Availability is a system performance parameter which provides insight into the probability that an item or system will be available to be committed to a specified requirement. Depending on the application, availability can be defined to include reliability, maintainability and logistic support information. For fleet management purposes, the ability to quantify availability in terms of all of its contributing elements is essential. This paper provides a discussion on a steady state operational availability model which can be used to assist the Canadian Air Force in its aircraft fleet management requirements. The availability model embodies scheduled and unscheduled maintenance and allows for impact analysis using in-service maintenance data. The model is sensitive to fleet size, aircraft flying rate, frequency of downing events, aircraft maintainability, scheduled inspection frequency and scheduled inspection duration. The predictive capability of this availability model is providing the Canadian Air Force with a more sophisticated maintenance analysis decision support capability. In order for this paper to be available for general distribution, it must be unclassified. As a result, the case studies presented do not reveal the actual operational availability of any Canadian Air Force fleet. However, the level of detail provided is more than adequate to illustrate the case studies and give insight into applications of the availability model
[http://ieeexplore.ieee.org/document/513274/]

The Frequency Distribution of Availability

Availability is probably the most informative indicator of performance for repairable devices. The construction and study of availability measures is usually focused on understanding the time evolution of device status-often relative to the efficiency of the repair process. It is not generally acknowledged that the resulting availability measure is actually an expected value with respect to frequency. At any point in time and with any associated value for availability, the number of copies of a device that is functioning is a random variable. The behavior of this random variable is the subject of the study described in the paper. The intuitive view that the frequency distribution of operating devices is binomial is confirmed. This is done using a combination of direct analysis and simulation for negative exponential and Weibull life distributions under the assumption of negative exponential repair time distributions. An efficient numerical strategy for computing availability in the Weibull case is constructed and provided as an ancillary result. The implications of the analysis are that more accurate models of device behavior in terms of frequency are defined including exact confidence bounds for availability at any point in time. In addition, the time evolution of the frequency distribution is described and the implications of this evolution for decision analysis are identified. The result of the analyses presented is a new perspective on availability that should prove quite useful.
[http://ieeexplore.ieee.org/document/291120/]

An Optimal Sparing Model for the Operational Availability to Approach the Inherent Availability

In an operating system, especially for the military, it is very desirable to achieve a high operational availability (A_O). In fact, the optimal goal would be for the operational availability to approach or equal the inherent availability (A_I). This could be accomplished by eliminating or reducing all time delays that keep the system down (nonoperational) and are not directly associated with the repair time (mean time to repair) of a failed item. The main culprit in keeping the system down is the availability of required spares. Therefore, if the logistic delay time due to the availability of spares is to be eliminated or reduced, the spares necessary to keep the system operational must be available at the location at all times. This, however, implies that the logistic delay time in the event of a system's failure, other than location delays, would be zero. Eliminating or reducing these delays requires an efficient and timely spares pipeline. An efficient and cost-effective way that this can be accomplished is to synchronize the spares pipeline with the system architecture and the fault-tolerant features. It is the goal for any system to be 100% available at the moment when it is required to accomplish a task. Therefore, an A_O optimization concept must be established. The maintenance and spare concept, in conjunction with an expert system using real-time computers as described in this paper, could make it possible to achieve the desired A_O

[http://ieeexplore.ieee.org/document/902476/]

Operational Availability Model of k-out-of-N System Under a Hard Time Maintenance Strategy

This paper considers a k-out-of-N hot standby system with identical, repairable components. A hard time maintenance strategy is used to maintain system. Failed components are replaced with spares by a component replacement group. Next, the replaced components are repaired by a different repair group. The system operational availability can be controlled by the system operating time, the spare part inventory level, and the number of repairman. We establish a mathematical model to analyze the effects of these variables on operational availability. Simulation results show that this model has the potential to optimize maintenance plan and logistics resource for k-out-of-N system.

[https://www.researchgate.net/publication/301254558_Operational_Availability_Model_of_k-out-of-N_System_Under_a_Hard_Time_Maintenance_Strategy]

A Study on the Case Study and Evaluation Methodology of Operational Availability for a Naval Ship using OT&E Data

Navy forces of Republic Of Korea (ROK) asked for more than 90% operational availability in the requirement document of combat ship. This study proposes the evaluation methodology of operational availability with the evaluation process, calculation formula, analysis of operational test data. As the case study, the developed methodology is proved to apply for 00 batch-I naval ship using the data to be acquired during the operational test period. The operational availability by test data was 90.03%, and it was satisfied with objective value 90%. The paper will contribute not only to establish the evaluation methodology of operational availability for combat ship but also other general weapon system.

[https://www.researchgate.net/publication/286897655_A_Study_on_the_Case_Study_and_Evaluation_Methodology_of_Operational_Availability_for_a_Naval_Ship_using_OTE_Data]

8.0 SUMMARY

Availability is considered as one of the most important reliability performance measures of the maintained system since it includes both the failure rates and repair rates of the system.. In today'scost centered industrial environment it has become very much essential to optimize the availability of the system to achieve maximum profit. Widespread demand for the availability improvement to an optimum point has emerged in so many areas like manufacturing, communication, power plant, airlines etc

Complex modern electronic and electro-mechanical systems have a huge number of critical failure modes, all of which must be addressed ideally at the design and concept stage, with the objective being to achieve a high level of Operational Availability . Usually these issues have to be addressed with finite resources. It is essential to carry out a formal reliability modelling exercise at the concept and development stages of any complex system.

At the stage of development, reliability engineers must inevitably rely on estimated failure rates and logistics delays based on assumptions of the proposed repair and support organization, in order to try to define the number and type of spare parts to be held forward. All of this has an impact on funding for the project taking into account such factors as redundancy, the initial spares buy and repair manpower requirements.

To achieve a high state of readiness or system availability requires that every element has been considered and estimated and that things like the initial spares provisioning, the repair organization, manpower levels and levels of training and test equipment are all adequate for the task. Even with all of the combined resources of design and reliability engineers, logistic planning staff and human resources specialists brought to bear on the task of estimating the need for support equipment, maintenance schedules, repair process definitions and required manpower levels, we are still left with the task of assessing every possible reason for system downtime. Only when these problems are addressed can we begin to put the essential elements in place to ensure that we have an adequate level of optimized Operational Availability..

9.0 GLOSSARY OF ACRONYMS

A_o	Operational Availability
ABC	Activity Based Costing
AAW	Anti-Air Warfare
ACAT	Acquisition Category
ACIM	Availability Centered Inventory Model
ADM	Acquisition Decision Memorandum
AIS	Automated Information System
ALDT	Administrative and Logistic Delay Time
AMSAA	United States Army Material Systems Analysis Activity
APB	Acquisition Program Baseline
APL	Allowance Parts List
AoA	Analysis of Alternatives
ARROWS	Aviation Retail Requirements Oriented to Weapons Replaceable Assemblies
ASUW	Anti-Surface Warfare
ASW	Anti-Submarine Warfare
ATE	Automatic Test Equipment
AW	Air Warfare
BCA	Business Case Analysis
BCS	Baseline Comparison System
BIT	Built-in Test
BITE	Built-in Test Equipment
CA	Criticality Analysis
CAIV	Cost as an Independent Variable
CASREP	Casualty Report
CEB	Chief of Naval Operations Executive Board
CINC	Commander in Chief
CLS	Contractor Logistics Support
CNO	Chief of Naval Operations
COTS	Commercial Off The Shelf
CRD	Capstone Requirements Document
DAB	Defense Acquisition Board
DID	Data Item Description
DOD	Department Of Defense
DON	Department Of Navy
DRMP	Design Reference Mission Profile
DT	Developmental Testing
DTC	Design to Cost
DT&E	Developmental Test and Evaluation
EDM	Engineering Development Model
FMECA	Failure Modes Effects and Criticality Analysis
FLSIP	Fleet Logistics Support Improvement Program
FOC	Full Operational Capability
FOM	Figure of Merit

FSC	Full Service Contracting
GAO	Government Accounting Office
GFE	Government Furnished Equipment
HALT	Highly Accelerated Life Testing
ICD	Initial Capabilities Document
IDE	Integrated Digital Environment
ILS	Integrated Logistics Support
ILSP	Integrated Logistics Support Plan
IMA	Intermediate Maintenance Activity
IOC	Initial Operational Capability
IOT&E	Initial Operational Test and Evaluation
ISEA	In-Service Engineering Agent
JCS	Joint Chiefs of Staff
JROC	Joint Required Operational Capability
KPP	Key Performance Parameter
LCC	Life Cycle Cost
LORA	Level of Repair Analysis
LSAR	Logistics Support Analysis Record
LRG	Logistics Review Group
LRIP	Low Rate Initial Production
$M_{Adm}DT$	Mean Administrative Delay Time
MAM	Maintenance Assistance Modules
MDA	Milestone Decision Authority
MDT	Mean Down Time
MDTD	Mean Down Time for Documentation
MDTOR	Mean Down Time for Other Reasons
MDTT	Mean Down Time for Training
MEC	Military Essentiality Code
MLDT	Mean Logistics Delay Time
MOADT	Mean Outside Assistance Delay Time
MOD-FSLIP	Modified Fleet Logistics Support Improvement Program
MOE	Measure of Effectiveness
MOP	Measures of Performance
MRDB	Material Readiness Database
MRTT	Mean Requisition Response Time
MSRT	Mean Supply Response Time
MTBF	Mean Time Between Failures
MTBM	Mean Time Between Maintenance
MTTR	Mean Time To Repair
NAVICP	Navy Inventory Control Point
NAVSUP	Naval Supply Systems Command
NSN	National Stock Number
OLSP	Operational Logistics Support Plan
OLSS	Operational Logistics Support Summary
O&M	Operations and Maintenance
OPEVAL	Operational Evaluation
OPTEMPO	Operations Tempo (pace of operations)
OPTEVFOR	Operational Test and Evaluation Force
ORD	Operational Requirements Document

O&S	Operating and Support
OT	Operational Testing
OT&E	Operational Test and Evaluation
PBL	Performance Based Logistics
POM	Program Objectives Memorandum
PPBS	Planning, Programming, and Budgeting System
PPS	Post Production Support
PRS	Provisioning Requirements Statement
PSICP	Program Support Inventory Control Point
PTD	Provisioning Technical Documentation
QFD	Quality Function Deployment
RBD	Reliability Block Diagram
RBS	Readiness Based Sparing
RCM	Reliability Centered Maintenance
RFI	Ready For Issue
RFP	Request for Proposal
RLA	Repair Level Analysis
R&M	Reliability and Maintainability
RM&A	Reliability, Maintainability and Availability
SAS	Supportability Analysis Summary
SMA	Supply Material Availability
SM&R	Source, Maintenance and Recoverability
SOW	Statement of Work
SRA	Shop Replaceable Assemblies
SYSCOM	Systems Command
SUW	Surface Warfare
SW	Submarine Warfare
TAAF	Test, Analyze and Fix
TAT	Turn Around Time
TECHEVAL	Technical Evaluation
T&E	Test and Evaluation
TEMP	Test and Evaluation Master Plan
ULSS	Users Logistics Support Summary
VAMOSC	Visibility and Management of Support Cost
VV&A	Validation, Verification & Accreditation
WBS	Work Breakdown Structure
WRA	Weapon Replaceable Assemblies

Design and Organize Manufactured Products for Optimum Operational Availability

10.0 ABOUT THE AUTHOR

Hilaire Ananda Perera has 40 Years of North American experience in Reliability/Maintainability/Safety Engineering
https://www.linkedin.com/in/hilaireperera/

- 2007 - Present: Reliability/Maintainability/Safety Consulting Engineer at Long Term Quality Assurance (LTQA), Markham, Ontario

- 1983 - 2007: Reliability/Maintainability/Safety Engineer at Honeywell Aerospace, Mississauga, Ontario

- 1981 - 1983: Reliability/Maintainability Engineer, Philips Electronics Ltd., Telecommunications Division, Scarborough, Ontario

- 1977 - 1980: Design/Reliability Engineer, DAF Indal Ltd., Mississauga, Ontario

- 1973 - 1976: Design Engineer, Stackpole Machinary Co., Scarborough, Ontario

Bachelor of Science Production Engineering (1972), University of Aston, Birmingham, England ## Professional Engineer (P.Eng) 1976 to Present – Association of Professional Engineers of Ontario ## Certified (1983 - 2007) Reliability Engineer – American Society for Quality ## Honeywell Six Sigma Plus Green Belt Certified (2001), Honeywell Design For Six Sigma Certified (2003) ## Allied-Signal Aerospace 1997 Growth Award

As a Senior Engineer applied mature engineering knowledge in planning and conducting reliability engineering and related product assurance projects. Emphasized the fact that reliability is the time-based concept of quality and reliability design is an iterative process that begins with specification of reliability goals consistent with cost and performance objectives.

Introduced new approaches to achieve high product reliability. Performed to satisfy internal and external customers. Developed awareness of reliability within the organization. Served as a technical advisor on reliability issues. Assisted management in using reliability information to make. decisions for.profitability.

SPECIALTIES: ** Use probabilistic design methods (Stress/Strength, Cumulative Damage) ** Parameter Mean & Variation calculation using Tolerance Statements ** Software (MathCad, SigmaPlot, FaultrEASE, Weibull) use for analysis ** Implemented Adaptive ESS to assure a very small Outgoing Defect Density ** Promoted RAC PRISM that use stress data and Process Grade Factors as a better than MIL-HDBK-217 ** Developed a model using Weibull Parameters to.calculate.Optimum.ESS.Thermal.Cycling.Time.with.Cost

PUBLISHED PAPERS and BOOKS:** Outgoing Reliability Assurance Using Chance Defective Exponential Model, AMMJ Mar 2015 ** An Introduction of Statistical Confidence Levels, AMMJ Sep 2014 ** Product Assurance Capability Quantified, Reliability Analysis Center (2Q2004) ** Adaptive ESS, Allied Signal Aerospace Company Technical Exchange Conference 25-26SEP1991 ** Optimum Cost Maintenance, Machine Design 20Jun1985 ** Reliability of Mechanical Parts, Machine Design 10Sep1987 ** Product Assurance ISBN 9781544843889 ** Adaptive Environmental Stress Screening Handbook ISBN 9781976014161